Taking the Plunge!

Taking the Plunge!

By
Bill McBride

BEST PUBLISHING COMPANY

Edited by Andy Nordhoff.
Layout by Dawña Argenbright.

Drawings by Steve Campbell.
Photographs by Bill McBride.

Printed and bound in the United States of America.

ISBN: 0-941332-87-X
Library of Congress Card Catalog Number: 99-069739

Best Publishing Company
2355 North Steves Boulevard
P.O. Box 30100
Flagstaff, Arizona 86003-0100, USA
Fax: 520.526.0370

Dedication

Commander Cousteau and the author.

Just as this book was being started, Commander Jacques-Yves Cousteau passed away in Monaco. Cousteau's energies and inventions literally "opened up" the underwater world to sport divers. Marine biology and underwater archaeology also benefitted enormously from Cousteau's pioneering. The last half of his life was devoted to the protection of our planet's oceans. To divers around the world, he was as the "patron saint" of diving and the underwater world. To government leaders, he was the voice of the oceans and of the oceans' inhabitants.

His voice, his efforts, his imagination, and his leadership will be sorely missed. The world is a better place because he was here. I respectfully dedicate this book to his memory.

TABLE OF CONTENTS

*Note: *The names of some of the individuals mentioned in this book have been changed to protect their privacy.*

PREFACE

"Every hour that one spends under water adds an hour to your lifetime!"

For the first few years of my scuba diving career, I quoted this modified fisherman's adage to many friends and students. The concept, of course, is that anything as truly RECREATE-ional as diving must surely be a "tonic" to life.

Later, I learned that the remark may have had far more validity than was ever guessed! Being under water while breathing air means that one is breathing compressed air. Breathing *compressed* air while the body is under the surrounding pressure of the water means that one is being exposed to an elevated oxygen partial pressure. The effects are exactly the same as if one was in a hyperbaric pressure chamber. At these higher pressures, the effect is the same as if extra oxygen is actually being "forced" into every cell in the body.

Medical technology has proven that this process can bring about a number of marvelously healing attributes. In a hyperbaric pressure chamber, the effect of gas gangrene can be reversed, for example, when the damaged cells are re-vitalized with the forced absorption of oxygen. Surgery can safely be performed on "blue babies" right in the chamber, when birth defects have affected the transfer of oxygen to their lungs. In fact, when the pressure is elevated enough, a person can almost survive without breathing, as the oxygen is driven directly into the body's cells without the need for the blood to carry it there.

As a side effect of experimentation in hyperbaric medicine, it was discovered that many other types of body cells can be regenerated with exposure to an elevated oxygen partial pressure — even brain cells! In many cases, ailing organs in the body can be given new life. One's energy level and sex drive can even be enhanced! A true FOUNTAIN OF YOUTH! Back in the 1930s, at least one totally pressurized sealed "hospital" was built (more like a hotel, actually), where patients (guests), could check in for two- or three-day visits, and leave with the feeling that years had been removed from their ages! Some wealthy individuals even have their own private, one-person chambers, for periodic "rejuvenation." (Michael Jackson has used one for years.) The book Uncertain Miracle describes the early history of this science.

So, it turns out that my philosophy was actually quite accurate! While the diver is under water, it is equivalent to spending time in a hyperbaric chamber! The more dives, the more "treatments." Being around active divers, one can't help noticing that they often tend to be more "energetic" people. Enthusiastic, effervescent, and ever ready for more adventures. And, it has often been observed, many of them also have an enhanced sexual energy! (It also happens that male divers father a higher-than-normal percentage of female children, though this has not yet been explained.)

At age 65, I certainly don't yet consider myself an "old" diver. Perhaps when I turn 80 I'll consider entering into geezerhood. In the meantime, I've just returned from a diving excursion to Micronesia, where we completed as many as five dives each day. In all modesty, my wife and I are often taken for 10-15 years younger, as are many of our good diving friends well into their seventies!

In fact, I have changed my adage to "Every hour spent under water adds at least a *week* to your lifetime! If that sounds like a "sales pitch," try it for yourself!

In any case, my 40 years of diving have certainly provided me with a lot of adventure, thousands of pleasant memories, and have brought me into contact with a lot of fascinating people. This book is about some of those adventures, and some of those characters....

1

We Take the Plunge!

According to a 1991 poll by the National Sporting Goods Association, scuba diving headed the list of sports that the most people "wanted to learn." This was from a list of 35 other active sports and hobbies.

In 1959, in northern Ohio, scuba diving wasn't on everyone's mind. There existed no Discovery Channel, Jacques Cousteau wasn't very public, and even *National Geographic* hadn't featured many articles on the underwater world. Lloyd Bridges was just starting the "Mike Nelson" television series.

The Aqua Lung had been invented (in France, by Cousteau and Gagnan) in about 1943. Mostly, the equipment had been used by "frog men" in the military. By around 1953, some civilian sport diving equipment was available in California, and soon *Skin Diver* magazine started publication there, mostly covering the exploits of spearfishermen. By the way, in 1953, when the French manufacturer asked the original U.S. importer of 12 sets of scuba equipment when he would be reordering the next shipment, he responded that the market in the States was now "saturated," and that he "doubted that any more could be sold." (Since then, more than seven million people have been trained and certified to scuba dive in the U.S. alone!)

Being outdoor-oriented, what little I could learn about underwater exploration caught my interest. It seemed like a new frontier.

Still, I never really quite put myself into this picture. It seemed, I guess, a more practical pursuit for an ex-Navy man or a west coast type, than for a midwest wholesale meat salesman.

Then one day when my Supervisor and I were calling on the meat manager in a Toledo supermarket, we were intrigued when the manager, Bob Diehl, mentioned that he was enrolled in a scuba class at the local YMCA, starting the next Wednesday night. "A scuba class?," we asked. Was it a Navy program, or what? Was he in the Navy Reserves, or something? "No," Bob explained, it was a "civilian" class, open to anyone with $25 and reasonable swimming ability. He explained that his own interest originated from the fact that he kept a motor cruiser on Lake Erie, and thought there might be some diving possibilities there. He went on to say that only three people were enrolled in the class, and that there was room for five or six more.

My supervisor, Bob Barton, and I drove directly from the supermarket to the YMCA and enrolled. Though we lived sixty miles from Toledo, the course was for only five sessions, and it seemed worth the drive. Bob and I, as well as being fellow employees, were also hunting and fishing buddies, and it didn't take either of us more than a few seconds to decide to get in on this new adventure! Another fellow employee, Dick Wilson, also signed on.

Bob was, indeed, the classic outdoorsman, always ready for a new adventure. His wife once gave him an authentic blowgun as a Christmas gift and, after practicing for several weeks, he brought it along on a rabbit hunting trip, leaving his shotgun at home. He didn't think he could hit a running rabbit with it, so we watched for a "sitter." When I spotted one in a clump of grass about 15 feet in front of us, Bob placed the needle-sharp steel dart into the blowgun and readied his aim. One sharp puff, and the needle struck the rabbit just above and behind the eye, with the other end thrusting out an inch or so on the other side of the rabbit's head. A brain shot! Right through the skull! Good aim, Bob! The red plastic plunger head on the dart kind of "twanged," like a quivering arrow, as it jutted from the rabbit's head. But the rabbit didn't twitch! Nor did it break and run! The one eye we could see remained open, and didn't glaze. Surely, it must have been an instantaneous painless death. What a shot!

Bob was overjoyed, and we were both exclaiming, as we walked up to examine Bob's first blowgun kill. As we stood before it, it seemed that it was still looking at us. As Bob reached down to it pick it up, it burst out of the clump of grass at full speed, and headed for a nearby woods! "Shoot it!" Bob said, but by the time I gathered my wits it was a rather long shot to make, with my sawed-off 20 gauge. A few pellets caught him, though, and the fur flew and the rabbit rolled head over heels. Then he got up and ran again! Eventually, we tracked him to a groundhog hole. There, I regret, is where he probably died. Apparently, the steel dart was so slender, and so sharp, that, like a hypodermic needle, it must have done little or no damage as it passed through the rabbit's brain. Bob never took his blowgun into the field again. If an archeologist digs up the skeleton of that bunny someday, he'll certainly have a puzzle to figure out!

Bob was a one-of-a-kind. You could hear his hearty laugh four rooms away, and it was more contagious than the yawns. He was the ultimate salesman, and probably could have made a living selling poison ivy seed or mosquito food. If you looked under "jolly" in the dictionary, you'd find a picture of Bob! Eventually, he even learned how to laugh under water. (It was a case of learn or drown!)

If he was midway on a day's schedule of sales calls and heard a good story, it was hard for him not to start back over with the morning's calls to spread that story back down the road. This was going to be a great guy to get into diving with!

At our first class, we met our instructor, Don Lee, Jr., and learned that a total of only six people were enrolled, including us and our friend, the meat manager, Bob Diehl. Don wasn't a certified instructor yet, but had been involved with the early stages of the YMCA Aquatic program, which had just begun to look into scuba training as another branch of its aquatic involvement.

We were surprised to learn that about half of our three-hour classes were to be spent in a classroom. We had assumed that the entire time would be spent in the pool. Soon it was clear that there was a good deal to know about the physics and physiology of diving, in order to understand the safety aspects and limitations of the sport. Though a textbook and standardized tests hadn't been developed yet, Don's notes and training were quite thorough, and we were required to pass certain written levels of testing. During one of our classroom sessions, someone asked the instructor how far one could expect to see under water. Don answered that in some of the very best of the diving areas, one could expect to see clearly for as far as 200 feet. "Wow!" said another student with thick glasses, "That's further than I can see clearly *above* water!"

In the pool, we had likewise expected to be using the scuba equipment from the very first. Instead, the first two pool sessions were spent in getting us adept and comfortable with the use of the mask, fins, and snorkel. We were a little disappointed, but later learned that this is the best approach. One of the early skills that Don insisted we perfect was the ability to clear the water from a flooded mask while under water. This is practiced by removing the mask under water, replacing it on the face, and exhaling through the nose while looking upward at about a 45 angle, while applying a slight pressure with the fingers to press the mask snugly against the forehead. Once mastered, this little trick is so easy that one could do it several times with one breath of air. But learning the technique is troublesome for many students. Bob, especially, had a really difficult time with this, and our sixty mile trips home from the first few lessons were with his eyes totally bloodshot, and his sinuses full of pool water. Neither of us were about to back down, but I must confess that at this point this whole thing didn't look like as much fun as we had thought it would be!

We persisted, however, and at the third session we finally got to breathe under water from a scuba unit. What a revelation! Completely submerged in water, and able to breathe pure dry air into one's lungs! By the time I took my third breath, I swore to myself that if I could find a way to do so, I was going to help introduce other people to this wonderful sensation. Weightlessness! Clear underwater vision! An ample supply of absolutely delicious air, and very little effort required to breathe it!

If I had never explored anything else under water but the bottom of that YMCA swimming pool, the five 120 mile round trips would have been worth it! Bob and Dick were ecstatic as well. After experiencing scuba, our trips back home from the classes were filled with plans to travel and dive, dive and travel. Of course, all of us had families, but we made grand plans for family vacations to areas where there were warm, clear waters. During the course, we became even better friends with the meat manager, Bob Diehl, who invited us to join him that summer on some dive cruises on Lake Erie on his boat.

Too soon, the classes ended. It was early March, and far too early to dive in any local lakes or quarries. We spent this time contemplating the equipment we would be getting, and figuring out where to get it. There weren't any dive shops in Toledo back then, or anywhere else in Ohio. Instructor Don had access to some "mail order" catalogs, though, so we met with him several times to select and order our diving equipment. Actually, there weren't that many choices. There were about two brands of scuba regulators and tanks, and only two brands and two types of wet suits. One type of wet suit was "ready made," and the other came in a "kit" form. Bob and I ordered the wet suit kits, thinking that we could tailor the suits for a more exact fit. The kits, in effect, were no more than a roll of sheet foam neoprene, a generic paper cutting pattern, a couple of cans of neoprene glue, and some zippers.

1961. Our first (yellow) wet suits were made from kits. Lead belt weights were cast in a large soup ladle.

Eventually, our gear began to arrive, piece by piece. Once we received the wet suit kits we had something to occupy our spare time for a while. We both had ordered yellow neoprene,instead of black, figuring that it would be easier to mark our cutting lines on the lighter color. It proved simpler than we expected to craft these "home-made" kits into rather presentable wet suits. And, in fact, they fit so well and looked so nice that they were to be later envied by other divers with "factory-made" suits! But then, all "factory-made" suits were black, so we *did* stand out a bit.

The making of our own wet suits led to one of our first diving "war stories," as a matter of fact. Five zippers came with each of the suits, and proved surprisingly easy to install. Each wrist and each leg was to have a zipper, to make it easier to get the suits on and off. (Nylon suit lining hadn't been "invented" yet, though, so one still had to use soap or talcum powder to make yourself and the suit slippery enough to slip on.) A longer zipper was supplied for the front of the wet suit jacket. Again, even this was rather simple to put in. Since each zipper allowed a little seepage of cold water into the suit, however, Bob decided to instead make his jacket as a "slip-over," with no zipper. With practice, he learned how to pull it on and off, much like a tight turtle-neck sweater. He tried to sell me on the idea, too, but I had broken both arms as a youngster, and had a bit of difficulty getting them in the proper positions to manage this pull-over technique. I went ahead and put the zipper in my jacket.

The third dive we made was in Lake Erie, from Bob Diehl's boat. It was a warm sunny spring day, and we each had two tanks, one for a morning dive, and one for afternoon. We were diving in an area that was sometimes used for military target practice, and we were looking for (and found) some spent munitions.

Exiting from the first dive, and intending to eat our lunch and take a bit of a break, we soon found we were getting overheated, with our wet suit jackets left on. I unzipped and removed mine, in short order, and Bob began rolling his up over his head. When he had it about half-way off, with his head inside the suit and his arms over his head, he dropped to his knees and started to grunt. I couldn't understand his muffled calls, but clearly assumed that he was "stuck," having trouble breathing, and was calling to me to help him get the jacket over his head. Quickly, I grabbed his jacket and began trying to tug it over his head and arms. He began hollering louder (screaming, actually), and fell thrashing to the deck.

I pulled harder. He screamed louder. Finally, I got the jacket to pop over his head, jerked it off his arms, threw it over to the side of the deck, and prepared to give Bob whatever further help he needed. At this point, he couldn't speak. His face was as red as a beet, and all he could do was sputter, make agonized faces, point at his wet suit, make gestures at his arm pits, and shake his fist at me! I was thoroughly puzzled. He obviously didn't need mouth-to-mouth, and I gathered that he wouldn't have taken it from me even if he needed it!

As he continued to point at his wet suit, which now lay in an inside-out heap on the deck, I took a closer look at it. Had there been some type of unknown creature that had crawled into his suit with him, and had been biting or stinging him? Maybe in his armpits? Then I looked a little closer. There, stuck to the inside of his suit jacket, right in both armpit areas, was about a half a handful of armpit hair! Everything suddenly became excruciatingly clear. By now, Bob was starting to get his voice back. He was using the foulest language I had ever heard him use, and he had an outstanding repertoire! I was almost ready to jump off the boat and swim to shore. What had happened was that when he was rolling his jacket over his head, he had rolled up his armpit hair in the sleeves. When he dropped to his knees and called for help, it was for me to roll the jacket back down, which he couldn't do, with his arms in that position. Instead, I pulled and tugged, jerked and yanked, and he couldn't make me understand that I was tearing his armpit hair out by the roots! (Try pulling out just one live hair!) We did make the second dive that day, but we didn't talk much for the rest of the trip.

That got me a little ahead of my story. While we were waiting for our equipment, and for spring and our first dives, the days seemed to take forever. By late April, we had received everything but our wet suit kits. A couple of divers from the Toledo area had told us about an unusual dive opportunity that was about to happen in early May. A local flooded limestone quarry was the site of an annual "trout derby." Each spring, two truckloads of adult rainbow trout were trucked into the quarry, and three days later a big trout derby was held.

During the first day or two after these trout were placed in the quarry, it was said that they tended to remain in a "school," until they got used to the new environment and dispersed. If a diver could manage to time himself into that quarry at exactly the right time, he might be treated to the unique underwater sight of a school of full-grown rainbow trout.

Being fishermen, Bob and I were intrigued by this opportunity. We learned the trout were being delivered to the quarry on a certain Saturday, and we could get there that very same day! However, our wet suits hadn't arrived yet, and rental suits simply weren't available. Nor were there enough divers around for there to be a chance of borrowing suits. We visited the quarry a week before the event, and our thermometer indicated that the surface water was about 55°F. Our fingers, immersed in the water, indicated that 55 degrees hurt. (Water carries heat away much faster than air, and will cool/chill an object, or a body, 900 times faster!) Still, we very much wanted to make this particular dive, or any dive, since we hadn't been in the water since our lessons ended.

We called the instructor, for his opinion. To dive, or not to dive, in 55 degree water, without a wet suit? He said "Don't do it, or it will be the shortest dive you will ever make!" We answered that we would likely try it anyhow, and asked if he could suggest ANYTHING that might help us stand the chill longer. He suggested that wearing something like long johns might help a little bit, perhaps even doubling our time, maybe allowing a four second dive, instead of two seconds.

Well, he obviously didn't know how determined we were. How used we were to hunting in cold winter weather. How often we had sat in cold duck blinds for hours. How much we wanted to dive!

We were at the quarry early that Saturday morning. It was a bright and sunny early May morning, with an air temperature of around 65 degrees. Beautiful! Two other divers were there (with wet suits), and told us that the water was exceptionally clear, and that, indeed, the trout were still circling the quarry in a tight school, and were a wonder to behold.

Our wives and our children, one by one, stuck their fingers in the water, looked at us, and found a place to sit and watch. We tested the water, too, and looked second thoughts at each other, but having hauled the families there, we had no choice but to dive. Suffice it to say that the instructor was right. It was agony getting our whole bodies into the water. No, we didn't do it the "painless" way, by jumping in. We inched in. By the time we got the last inch in, the first inch was done for. We breathed from our regulators, ducked under water, looked at each other, and looked down at the bottom, about 40 feet below. Our families were watching. Our children were watching. Our boys were watching! We had no choice. We had to at least disappear from sight. We nodded at each other, gathered our determination, and headed for the bottom. We got there quickly. At the same time, we got to our limit of resolution. We quickly headed for the surface, climbed out, un-geared, toweled off, and immediately decided to take a walk to the other side of the quarry, where we shivered uncontrollably, out of sight of our families, for the next thirty minutes.

Needless to say, we didn't get to see the trout (though we did the following year and even got some pictures of them). We dutifully entered this first dive in our log books, but I noticed some years later that someone(?) had written in a zero after the single-digit number of minutes comprising our "down time" on that dive. We also later learned that, in this part of the country, nearly all lakes and quarries are about that same 55 degrees in the bottom layer of water, even in the summer time when the surface water is far warmer. We learned, too, that a well-fitting wet suit can nicely handle this temperature, and keep the diver reasonably comfortable. But what we learned *that* day was "long johns don't substitute for wet suits!"

We often returned to that lake, which was a flooded, abandoned limestone quarry just west of Toledo. For a freshwater quarry, it contained a number of interesting things. One was a large safe, with its door obviously having been violently removed and laying nearby. Almost certainly, the result of a theft.

In fact, we quickly became aware that almost any body of water which had shores accessible to man was likely to contain "loot," or evidence of this or that crime. This was particularly true of areas where a bridge crossed the water, or a road was near the shore. It seems that it is human nature to consider the surface of any water to be a membrane to an invisible world. At least until sport diving became popular, the best way to discard anything was to throw it into water — particularly if it was something it was hoped would never been seen again! Having realized this, we relished any opportunity we could garner, to dive in such places that had never been yet visited by divers. Back then, there were many such lakes, ponds, and quarries, and there were countless "relics" to be found. Guns of all types, stolen goods which became too "hot" to deal with, bicycles, motorcycles, cars, more safes, and trunkloads of all types of such things. Most such "finds" were made worthless by the immersion in the water, but now and then something was found to have salvage or "souvenir" value. Now and then even bodies were discovered, but thankfully not many by us. The classic such discovery in northern Ohio occurred when a pair of divers came across a zipped-up bedroll, 80 feet under water at the base of a sheer quarry cliff. Upon un-zipping the bag, they found the body of a middle-aged man, accompanied by a number of large rocks. Apparently, this guy was camped near the edge of the cliff, and had warmed some rocks by his fire, to stuff into his bed roll to keep him warm, then accidentally rolled of the edge of the cliff and into the lake. Sure....

Bob and I never found anything of great monetary value under water. One of the more interesting finds, though, was a chain saw. Laying among a tumble of rocks at about 60 feet of depth near the edge of a quarry which we dived regularly was a lump of limestone accretion in the shape of a chainsaw. We never bothered salvaging it because it was so obviously ruined. After pointing it out to other diving buddies on several dives, I was moved one day to wrestle it up the side of the quarry and onto the shore. Later, throwing it into my trunk, I hauled it home and carried it into my basement.

Months later, in the off-season, I decided to see if the thing might somehow have any salvage value. As I started to remove the limestone crust, I noticed the remains of what appeared to be a service tag attached to the handle. Before the evening was over, I had all of the crust off of the saw, and was surprised to see that it looked to be quite intact! Without adding a single new part, not even a new sparkplug, I eventually got that saw running and usable!

Contact with the local police later indicated that the saw was one of many that had been stolen from a local repair shop *three years before*, and never recovered. All of those stolen were from the shop's shelf of completed jobs. Apparently, the one I found had been thusly serviced, and so thoroughly oiled and lubricated that it was able to withstand three years of immersion. Though its paint and some of its thinner sheet metal was affected, all of its moving parts were well preserved. The police told me to just keep it. Eventually, I wrote to the original manufacturer, thinking they might like to hear this story of the "durability" of their products (and just might, hopefully, offer me a complimentary brand-new one, but they only offered to provide any parts I may need for it).

Another time, in another lake, we found an intact wooden crate, still nailed shut, of a size that might be expected to hold something like maybe about a half dozen rifles. It was about the right weight, too, and we thought we really had something here! We couldn't get the crate open while under water though, so we muscled it up the slope and onto shore and removed the lid there. Instead of rifles, we found about 200 pounds of various sizes and shapes of what appeared to be scrap metal. Disappointed, we took away a few of the pieces that we thought might be useful around home for whatever, and threw the rest back into the water. Years later, I mentioned the incident to a machinist, and he told me that the crate was probably full of a salesman's samples of tool quality steel, which can be quite expensive. I dug up one of the pieces of "scrap" that I still had kicking around, had him analyze it, and we learned that this was actually the case. However, the quarry where we found it had since been filled in.

Back to that ordnance we had found in Lake Erie. We made several dives in this area that was a designated target area for Camp Perry, a lakeside military base which was basically an ordnance disposal facility for the military. This particular area of the lake was closed to the public on weekdays, and tons of out-dated or experimental munitions were fired into the lake at a small island, or at buoyed targets, or simply at random. Some of the out-dated smaller ammunition, grenades, etc., were stuffed into the hollow explosive chambers on some of the big howitzer shells, from time to time, and just "expended" into the lake. Eventually, we recovered all kinds and sizes of ordnance, including many grenades and mortar rounds, and hauled them home. Some of the spent bullets, mostly 50mm, had some value as scrap metal, and these could be collected by the bushel. The rest, we just stowed around the house or the yard. Eventually, we got around to asking a munitions expert to look them over. He quickly (and nervously) advised us that much of what we had collected and hauled home was still "live," and should be considered quite dangerous! What to do with it? Since no one at that time was in the business of disposing of such, he could only suggest that we transport it back to the lake and dump it back where we found it. Eventually, we did exactly this, and considered ourselves pretty lucky!

2

Diving for Dollar$

Bob and I, occasionally joined by Dick Wilson, made many dives that first summer. Eventually, we learned that another individual from our home town was also into diving. Ron Heredon, a un-married "free spirit," had bought some home-made equipment from a friend in the Air Force, and had "learned how to dive" from some other friend, somewhere. His breathing apparatus was made-over from an aircraft oxygen mask, and his tank was an old CO_2 bottle. He was attracted to us because we had a means of getting our air cylinders refilled. (By taking turns transporting them all the way to Toledo, to the only high pressure compressor that then existed in that part of Ohio.)

In general, we tended to steer clear of Heredon, as we sensed that his knowledge of diving safety was sketchy, at best, and we had been warned to steer clear of make-shift scuba equipment. He was a nice-enough guy, and we didn't want to seem too "sophisticated," but he seemed a bit reckless, compared to the way we had been trained to approach this sport.

Actually, Ron did do one swell thing for me. Our home town, Defiance, is located where two rivers meet, the Auglaize and the Maumee. Both are far too muddy to provide good diving, except for the occasional "grope" under one or another bridge, to find what "treasures" may have been clandestinely thrown into the water.

Since the town was built on both sides of both rivers, several water and sewer lines passed under the waters. Ron worked part time for the water department, and it developed that a substantial leak was suspected to exist in a major water main that passed under the Auglaize. He was hired, on the side, to locate and repair the leak, and asked me to join him in the task. In the end, the revenue from this one job paid for all of my initial diving equipment.

First, from the river's shore, we followed the eighteen-inch metal pipe across the riverbed, to locate the leak. This proved easy, even with no visibility. As soon as our heads were submerged, we could hear the high pressure leak "singing." As we got closer to the leak, the noise grew louder. It proved to be a faulty joint, only about 30 feet out from the west bank, in a depth of only about twelve feet. So much water was leaking that an area of several feet around the joint provided fine visibility, once illuminated with our underwater lights. The volume of the leak was such that the water department estimated they were losing some 100,000 gallons of treated water per day!

Finding the problem, of course, was one thing. Fixing it was another! The leaking joint was of the bolted flange type. Twelve two-foot bolts held the flanges together. We were provided with large wrenches, to tighten the bolts. Indeed, we found a few of them to be somewhat loose. However, once all of the bolts were as tight as we could get them, the leak was only slightly diminished. "No problem," said the engineers. If we could excavate a space in the river bottom under the flange, a halved, gasketed "joint splint" could be fabricated, which we could bolt over the entire leaking joint, thusly containing and stopping the leak. This sounded easy, but once we learned that each half of the "patch" would weigh nearly 150 pounds, we realized that a floating working platform would need to be provided. The City water Department had no such barges.

We rented a home-made pontoon boat from a local fisherman, and obtained a 20-foot ladder. Tying the boat off onto neighboring joints in the water main, we then secured heavy weights to the bottom of the ladder. Now, we had a direct and easy route from the surface to the leak. Weighting ourselves quite heavy, we transported each half of the new patch to the faulty joint with little trouble. By the end of the day, we had the patch bolted on and the leak stopped. As the last of the bolts were tightened the sound of the leak diminished, then stopped altogether. What a nice silence!

A photo of Ron and I in our dive gear on the "work boat" appeared on the front page of the next day's local paper. This eventually led to a few other contacts for local underwater work, but "commercial diving" wasn't what any of us really had in mind. The proceeds from this one job, however, did ease any feelings of guilt that I may have felt about spending too much money for my equipment.

In the years since, dozens of my students, totally enthralled with diving, have inquired about the possibility of "making a living" by diving. I have always felt obliged to counsel them that the job of "commercial diving" is as different from sport diving as play is to work. The high-paying jobs in commercial diving are actually skilled construction or mechanical jobs, the only difference being that they are conducted under water. In fact, most of the companies that employ commercial divers usually say that it would be far easier to teach a good mechanic to dive than it would be to teach a diver how to be a good mechanic. In other words, one would have to really love hard, dangerous, physical labor to "enjoy" commercial diving! And, one's chances of finding a job as a commercial diver would be slim, indeed, unless he or she was already a skilled welder, pipe-fitter, demolition expert, etc. And, the conditions under which a commercial diver is usually called on to work are absolutely nothing like the sport diver experiences on a pleasure dive.

To have the slightest chance of becoming a commercial diver, one not only needs to have an already established skill in one or another mechanical trade, but a great deal of expensive training is involved, in the use of the specialized equipment required. Though several "commercial diving schools" now exist, the vast majority of their graduates do not proceed into this career. (In fact, it is said by some that there is more profit in running these schools than there is in the diving itself!) Those who do stick with it, enticed by the fact that commercial divers can easily make over $100 per hour, find that they must first serve a term as another diver's helper, or "tender," at far more modest wages. This phase can often last for years, before the candidate actually makes his own first dive. Then, once finally diving, the individual may well find himself employed for two months straight (perhaps all at sea), with little or no time off, then un-employed for several months at a time.

Commercial diving is not a glamorous career, by any means, and the way it is conducted, often under economic stress by the employer, it is often a very hazardous job. Only two of my former students have chosen to go this route. One is still at it. His work is here in the midwest, all in municipal wells, the Ohio river, sewerage systems, water towers, etc. The other fellow actually went into the "big time," working from oil-drilling platforms off-shore of New Orleans. He gave it up after about four years. Though he made BIG money at the times he was actually diving, his total annual wages proved to be little more than a typical construction worker. And, though he was lucky himself, he reports that many, many of his fellow divers suffered injuries or diving-related maladies, some serious, and few of them adequately compensated for. If you remain with any doubts about such observations, check with your life insurance agent as to the going rates for commercial divers!

On the other hand, I've known many sport divers who, like I did, have sometimes taken on minor part-time work under water. Often, such jobs involve locating something someone has lost into the water, like a valuable duck hunter's gun, or such. Usually, such jobs are contracted on a "no find, no pay" basis. Others find some occasional work repairing pools, pond outlets, docks, etc., or cleaning the bottoms of boats. Under the right conditions, such jobs can be lucrative, though I suggest that a written agreement be signed, prior to the work. Two friends of mine once found and raised an $1,800 outboard motor for a boater, who then offered them $20 for the job. THEY THREW IT BACK INTO THE LAKE, and told him to get it himself!

One fascinating "underwater job" we took on was the inspection of an elevated water tower. The city of Weirton, West Virginia, hired myself and my friend, Jack Porter, to climb, enter, and photograph the interior of a large water storage tower. The tank was about 2/3 full of water, as local insurance codes would not allow it to be drained for inspection. It was 135 feet from the ground to the bottom of the globe-shaped tank, which was several years old. During construction, it had been equipped with a number of electrolytic zinc anode rods which were expected to retard oxidation on the interior walls of the tank. The city engineer wanted photographs of how well these rods were doing their job. In the end, we found that the rods had done their job quite well, but only within a limited area of each one, and that a good deal of serious corrosion existed elsewhere. We got some great photos, and the engineer was quite pleased with our work.

While the results of this job were rewarding, actually doing it was far more complicated than we had anticipated. We had to climb a steel ladder all the way to the catwalk that encircled the tank, then clamber up and over the top to the entry access hatch. I climbed first, with a coil of rope, then hauled our equipment up, piece by piece, tanks, wet suits, and all. That was a job by itself. When Jack arrived and we entered the water, we were pleased to find it quite clear, as we had hoped, and of a temperature that was reasonably comfortable. The open hatch surprisingly provided enough light to illuminate the entire interior, so we hardly needed to use our underwater lights. It was rather eerie, when on the bottom of this tank, at a water depth of about 20 feet, to think that the "bottom" consisted of about a half inch of steel, and that under that steel was about 135 feet of air.

Once our inspection and photographs were complete, we exited the hatch and removed our gear on the catwalk. I descended first, and Jack lowered our gear on the rope. We decided that he could simply drop the unbreakable items, such as wet suits and fins. Unthinking, I attempted to catch the first fin he dropped. With luck, it "sailed" away from me a bit, and I missed it. Good thing! That fin hit the ground hard enough to break one's skull! Ka-WHUMP! Those old guys Newton and Galileo obviously knew what they were talking about when it comes to gravity! From that height, everything he dropped really gathered velocity, and yet another of life's little lessons was learned....

A more realistic diving-oriented career exists in the field of sport diver education. As the sport grows, more dive shops need more instructors, and a number of full-time jobs are almost always available. I personally have found it a very rewarding enterprise, and have made countless new friends in the process. It's more than neat, I think, to be around people who are really excited about what they are into! Several instructor training and accreditation schools now exist, all around the country. A "side benefit" of a career in this field is that instructors often have the opportunity to serve as paid trip escorts for the dive shop's group trips to many exotic dive destinations.

Another career that intrigues many diving enthusiasts is the prospect of finding employment as the "divemaster" at one of the dive resorts located in several "tropical paradise" locations. At first thought, this seems the ideal vocation. There are, however, a number of serious drawbacks. First is the fact that so many eager candidates are vying for these jobs that the employers have found that they don't need to offer much pay. In fact, some of these dive guides work for little more than room and board. And, since the actual time that clients are actually diving represents only a fraction of the time, those guys and gals with this glamorous-sounding occupation find themselves called on to do a great deal of the "drudge work" associated with the operation of a resort, or diving charter boat. Boat repair, tank filling, building maintenance, and a lot of gear lugging are often considered a part of this job. And, invariably, a certain percentage of the diving guests turn out to be real "lemons," and require undue amounts of baby-sitting and hand-holding.

In actuality, most folks "burn out" pretty quickly in this type of operation, and the turn-over rate is horrendous. This seldom is a problem for the resort, however, because whenever a dive guide leaves, a dozen others are eager to take his or her place! Then, too, in the case of many foreign countries, the resort owner is required to employ only native residents, and "outsiders," when they are fit into the operation, find themselves working virtually as volunteers, or on some sort of quasi-legal basis.

If you've ever been to a diving resort, you likely felt that your dive guides had found a true "dream job." Seemingly carefree, diving most every day, usually most gregarious, spending "social time" with your group, and living full-time in the tropics — what more could a diver want? Your vacation ends and you leave the scene and go back to work. These folks, it might seem to you, get to stay on vacation! What a life! Look closer, however, and you would often find that these "lucky" individuals are paid little, have few, if any, employment benefits, and have a heck of a lot of hard work to do when you, or the next group, aren't diving. Certainly, this life style fits perfectly for some types of people with a vagabond spirit and few financial responsibilities. But, in most instances, it is definitely not a life-time career with much future!

3

We Explore Our New World

During our first few years of diving, we often got calls from individuals who were interested in having us "check out" this or that lake or pond that had never before been dived.

A pattern developed in such calls. We would be assured that the water would be "crystal clear," and, in many cases, that the depth of the water was "bottomless," or "at least 70 to 100 feet!" And, often, that legends existed that indicated all types of "booty" had been hidden in the lake.

We quickly learned that "crystal clear" meant that a tumbler of water dipped from the lake could be "seen clear through." That is, a glass full of the water did, indeed, look rather clean. However, in all but the rock-shored flooded abandoned limestone quarries, the actual underwater visibility usually turned out to be almost zilch. Even if a few inches of visibility did exist at the surface, a state of sheer blackness (or extreme brownness), was all that remained at a few feet of depth. In most cases, if anything was on the bottom, it was going to have to be felt, not seen. Spooky, indeed, and not the kind of diving that one finds exhilarating! And, we quickly learned that a water grave was not only appropriate for stolen "loot," weapons, and other hardware that someone wanted to make "disappear," but that lakes and ponds were also a favored dumping ground for trash, discarded rolls of fencing, junk cars, and even deceased farm animals.

And, as to the countless "bottomless" lakes and ponds we were called on to explore, the vast majority were far more shallow than the owner had judged. Farm ponds, particularly, and especially if they are fed by a stream, tend to "silt up" in a very few years. In some cases, the owner had built the pond, and knew exactly how deep it was when it was new. Perhaps, for example, a pond was originally 30 feet deep (which is a good depth for keeping the water cool and weed-free). Ten years later, though the water level remains the same, that same pond may well be silted in to a depth of 6-8 feet. The 20 feet or so of silt that has been carried in to the pond is so soft and "gooey," however, that it is, indeed, hard to locate the "bottom" with a weighted line.

Divers submerging in to this semi-suspended silt sometimes find themselves "swimming" through the blackness of the liquid mud. Not a pleasant place to be...

Eventually, we learned to respond to such invitations to explore this or that "diving paradise" with a good deal of misgiving. And, despite the often-related legends of submerged booty and treasure, we seldom found anything of any real value in such places. Still, it was an adventure to be the very first to explore the bottom of a body of water, and we seldom turned down an invitation!

We did, early on, in a flooded 60-foot-deep rock quarry, find a quite intact 12-foot metal rowboat. With no flotation chambers built in, it had probably been accidentally sunk. Turning it upside-down and exhaling our exhaust bubbles into it, we were able to float it to the surface and haul it out on the bank. Later that same day, we were snorkeling around on the surface without our scuba gear, when one of our party launched the boat and was paddling around the quarry in it.

As a joke, I swam up behind the boat, and put all of my weight on the stern, causing it to flood and sink, leaving the "boater" on the surface, boat-less. As the boat glided toward the bottom, it suddenly occurred to me that some other scuba divers may still be on the bottom, and that this boat could land on their heads! Taking a large inhalation of air, I caught up with the boat, intending to "steer" it away from any divers who may be under it. At a depth of about 30 feet, I could see that no divers were in its path, and, at the same time, I realized that I was at the limit of my breath-holding ability! I released my grip on the boat, and angled toward the surface. However, just as they say about sinking ships, this little rowboat, as it descended, was creating a powerful "down-draft" of water, which had me in its grip! Try as I may, I couldn't fin out of the boat's slip-stream. I was pulled deeper and deeper, until, finally, as the boat approached the 60-foot bottom and began to level off, I was able to break free. By the time I reached the surface, I was on the edge of unconsciousness. This was my first "close call," and would have made for a really embarrassing obituary!

Later, an intact Ford Model "T" coupe was found on the bottom of this same quarry. As a publicity stunt, a local diving club announced that they would stage a public demonstration of how a car can be raised from deep water, with the use of flotation bags. A date and time was published, and, when club members arrived at the quarry that morning, the Model T was found sitting high and dry on the shore, compliments of a competitive club that had done the job in the middle of the previous night. On the windshield they had attached a note: "If you want to get it any higher, you'll have to use helium!" Nothing for the club to do, then, but to push the car back into the water, and raise it again when the crowd arrived!

This same Model T later proved to be made memorable to a number of the students of a certain local instructor. Before taking his new students into this quarry for their first open-water dive, he would describe the antique car they were likely to spot. Finding it, he would hold back and signal his students to go ahead and explore it. Typically, they would go right for the steering wheel, to see if it still turned the wheels (which it did). They'd open the doors and swim in and out of the car. Eventually, they would move around to the rear of the car, intending to see if the trunk still opened. The instructor happened to be a part-time amateur magician, and, among his props, had a number of realistic human "body parts." He delighted in having previously placed a "human" arm or leg inside of the trunk, partially obscured by some debris.

As the new students proceeded to the rear of the car, found that the trunk lid would open, and suddenly focused on the fact that there appeared to be a human body in the trunk, they would usually react in ways that delighted the instructor. In fact, some of them would lose it completely, and he would have to go to their "rescue," which, he felt, added to their training experience. In reality, a number of his students did not appreciate the "joke" at all, and thereafter had a much lower opinion of the instructor's sense of humor. Others swore to get "revenge" on him, sometimes succeeding in some rather ingenious ways!

Our equipment, in those early days, was rather primitive by today's standards. We made our own belt weights by melting down scrap lead and pouring it into soup ladles, then cutting the belt slots with a chisel. Inflatable life vests were just starting to become available. Our own first "emergency ascent devices" were rubber balloon-shaped floatation devices inflated by a small CO_2 seltzer cartridge. These were worn at the waist, and, in a real emergency, would have provided little, if any, salvation. (When yoke-type life vests later became popular, these earlier "balloons" were then marketed as "marker buoys," a function they served quite well.) Mask bodies were made of rubber (instead of today's silicone compounds), and usually deteriorated after only a couple of seasons, often leaving a ring of gooey rubber on one's face. We fashioned our own "diving knives," the question often debated being, "should a lost knife float, or sink?" That is, was it better to go to the surface to recover a loose knife, or have it fall to the bottom? Our cylinders were attached to our backs with an arrangement of nylon straps fastened to ring clamps on the tanks. In use, they would often float over one's head, swing around under the arm, or, most often, be dangling in the area of the ear.

Soon after our entry into the sport rigid "back-packs" for the cylinders were developed, and this greatly added to the diver's comfort. Bob and I had chosen to have our tanks equipped with "J" valves. This valve automatically shut off the diver's air supply when the cylinder pressure was depleted to about 300 psi. When a lever on the valve was activated with a pull rod, the last of the air became available to allow the diver to surface safely. The alternative to the "J" valve was the "K" valve, which had no reserve feature. No one, however, could tell us what the initials "J" and "K" stood for. Eventually, we learned from a manufacturer's engineer that the designations were simply derived from the fact that, in the country's very first diving equipment catalog, the reserve-type valve was catalog item "J," and the non-reserve valve was item "K."

Speaking of buoyancy, it once occurred to Bob and I that an "artificial" method of creating underwater lift might be the "effervescent" principle. Noting that fizzing tablets, like Alka-Seltzer®, seem to create a lot of gas bubbles when placed in a tumbler of water, we decided to see what they would do under water. Carefully placing about a dozen such tablets in a plastic waterproof bag, we proceeded to a depth of about 30 feet, where we positioned a coated canvas bag, upside down, to catch the bubbles that would result. Opening the water-tight bag, we exposed the tablets to the water. They fizzed, just like they would in a tumbler. However, the amount of gas that resulted from the dozen tablets, all of which was trapped in our "lift bag," amounted to not much more than a few cubic inches — hardly enough to float the bag itself. So much for that experiment! I've often wondered, though, if some type of "super-effervescent" powders or tablets might exist, which actually could make enough gas, under water, to be of some use, perhaps as an emergency inflation device....

Underwater photography, which we all became interested in, was a particularly wide-open field for innovation, since no truly water-proof cameras were yet being marketed. We experimented with cameras in jars, cameras in plastic bags, and, eventually, land cameras in home-made clear plastic housings. Suffice it to say that we "drowned" many a camera! When we eventually advanced to the desire for better underwater exposures, we began to "invent" means to introduce the use of flash bulbs to our underwater photography. For reflectors, we often used large soup ladles with a hole drilled in the center for the bulb socket.

The difficult part of this was devising ways to connect the camera, inside the housing, to the flash unit on the outside, without creating a leak in the housing. And, it was always a hassle to control the supply of new and used flash bulbs, which floated. Often, we would simply put the bulbs in a sock, which then floated upward from a cord attached to the diver, or stick them into small holes punched into a strip of inner tube rubber. At first, we would simply release the used bulbs, which would float to the surface, but we soon realized that this practice was not couth, and instead collected them in another sock, for proper disposal. We once saw a large bass strike and swallow a used flash bulb as it floated to the surface, but he shortly disgorged it. (He didn't want a light lunch, apparently.)

One of the author's early "home-made" camera housings.

I became fascinated with the idea of underwater movie photography, and built several Plexiglas® "prototype" housings for various 8mm cameras. Eventually, I learned how to construct housings which actually did work, without leaking. My designs proved so practical that several other divers had me make housings for their cameras. The field of underwater 8mm movie filming, though, did provide a real challenge, particularly because one had only 90 seconds of filming on each side of the three-minute film roll. Once the first half of the film was used, one had to surface, remove the camera from the housing, reverse the film, and start over. Then, of course, there was the week or so of wait, while the film was developed, before the results were seen. What a far cry from today's underwater video outfits, providing up to 90 minutes of filming, and instant viewing!

About the same time we got into diving, a few small scuba clubs started to be formed, in some of the larger midwest cities. Once some of these clubs made contact with one another, some underwater "competitive games" were invented, to provide an opportunity for the clubs to interact. This was already happening, of course, on the west coast, where the competitions usually involved spear-fishing events. Since spear-fishing was basically illegal here in the midwest, other "events" were devised. Clubs would come together at one quarry or another, in the summer months, and stage measured and timed events such as the "compass course," "find the lost object" games, "team buddy breathing," "equipment exchange," and so forth. Bob and I were permitted to participate in some of these events as free agents, since we had no club affiliations. Thusly, we made a number of new friends in the diving community (and even won a few trophies!). Some of our new acquaintances proved to be life-long friends, and some were later invaluable in the career course that eventually developed for myself.

Naturally, as each new underwater game was thought up, all of the clubs and divers would immediately begin to practice the exercise back at their own home water. Some would concentrate on thinking up ways to "out-smart" the rules. There is, it seems, a rather fine line between *innovation* and *cheating*. And, divers being the inventive and adventuresome types they are, the outdoor competitions often became quite lively events! Usually, the day would end with a mass "treasure hunt," with all of the divers searching for numbered markers which would be redeemed for prizes provided by local merchants or some of the diving equipment manufacturers. Of course, diving competition, like submarine racing, isn't much of a spectator sport. In general, however, the day proved to be as much a "social" event as a competition, with most of the divers bringing their wives and families for a group picnic and camp fire party. Ninety-nine percent of the divers I've met are gregarious and convivial types who seem to have an unusual zest for life and are fun to be around.

Speaking of diver's personalities, I've often been asked if I could identify any personality generalizations between those individuals that discover diving and adopt it as a life sport, and those who simply give it a "try," and drop out. After 40 years of observation, I'm convinced that certain factors do seem to hold true. First, there are a number of different reasons why individuals first get in to scuba diving. I'll not say there are "right" or "wrong" reasons, but there seem to be certain motivations that, if identified, would be seen to most often lead to one path or the other. When sport diving was young, it was seen as much more of a "macho" thing than it is today. Certainly, there were many guys who started diving because it seemed a really "man" thing to do. That factor still exists, to some degree today, even though a growing number of divers are female. Those who enter the sport basically to "prove" something tend to be far less likely to continue with it. In fact, bring up the subject of scuba diving in any crowd of people, and several folks will readily volunteer that "Yeah — I scuba dive!" (or "*used* to scuba dive," etc.). Their manner will often infer that they have easily "mastered" that particular sport, then moved on beyond it to other things. Often, for these types, the initial goal seemed to be to simply file another "bragging right" in their repertoire of conversation.

Of course, there are other reasons why an individual who has completed a scuba course might choose to "drop out." Clearly, there are those who think they will enjoy it, and simply learn that they don't after all, after a few open water experiences. As different and as exciting as it is, it isn't for everyone. Sadly, this conclusion is often reached because the individual's early diving experiences have been inadequately planned, inappropriately supervised, or conducted under unpleasant conditions. Though everyone can't complete a scuba course and head immediately for warm, clear, tropical coral reef diving, there is little point in exposing oneself to early experiences which are uncomfortable, un-enjoyable, and unduly stressful. Though one can expect to be a bit apprehensive during the first few diving experiences, it is self-defeating to place oneself in truly negative conditions. This, by the way, is an excellent argument for joining a dive club, and/or enrolling in further "advanced" diving courses.

And, there are those who are coaxed, coerced, or literally "forced" into a scuba class, as by a spouse or a friend determined to create a "handy" diving buddy. While a few such folks might actually get intrigued by the sport, the vast majority will do poorly during the training, participate reluctantly, and never be "turned on."

Some individuals have an inherent fear of the water, and even if they can will themselves through a scuba course, they will likely never be at all comfortable in the sport. Those who really get into diving tend to want to get everyone else they know involved. That's understandable, but one should never "muscle" anyone in to it who indicates a negative feeling for it!

In those who do continue the enjoyment of diving, I can see a number of personality similarities. Above all, the real enthusiasts, and I know hundreds, at ages ranging up to 75 and beyond, all share a bright and shining sense of CURIOSITY, or awareness, about the world around them. These folks will surface from their 800th dive exclaiming "WOW! Did anyone else see that little red and yellow fish by the purple sea fan?!?" They anticipate dive trips as a child anticipates Christmas, and they continuously find wonder in every dive. While the "temporary" diver might say something like "Yeah, I dived that lake once, no sense going back there," the life-sport diver will delight in diving the same area over and over, seeing new things, sensing seasonal changes, fascinated by the differences of the world beneath the surface. Childlike? Perhaps. But these folks seem to have no selfish motivation beyond the enjoyment and appreciation of the wonders of nature. They seem, above all else, to be connected to the natural world!

I'll mention one other category of enthusiast. This is the diver who, after the first several dozen dives, begins to lose some of the enthusiasm for diving. Not exactly "jaded," this individual does gradually tend to see less to be excited about. Then, by one means or another, he or she gets interested in underwater photography. It's true that the diver does have difficulty in verbally describing the underwater world to friends who have never experienced it. Now, with slides, photos, movies, or videos, the diver can bring some of this "new" world to the vision of others. With the camera, the photographer looks "closer" for subjects. Where he was "seeing all the bottom he could" on each dive, before, now he is slowing down, and seeing far more! A coral reef, for instance, to a passing diver, can appear devoid of fish. Once the diver stops and remains stationary for only a few seconds, however, the entire reef formation comes alive with hundreds of life forms that had been frightened into countless crevices by the moving diver.

This is especially true if the photographer specializes in close-up work. A whole *new* world is opened when the smallest of areas is scrutinized for subjects — even in a freshwater lake! In fact, today, if I am on a diving excursion with, say, twelve other "mature" divers, I would expect at least ten of them to be underwater amateur photographers, with at least eight of them equipped with close-up equipment! Though I here used the word "amateur," many of these photo-hobbyist's work will equal the quality of publications such as *National Geographic!*

I've often been asked, too, how sport scuba diving is that much different than the myriad other hobbies available. This one isn't hard to answer. Lots of other sports and hobbies provide a great deal of enjoyment, recreation, and challenge. In fact, many divers, like myself, also enjoy other sports as well. Some of these, especially the outdoor sports, can be quite healthy and fitness-enhancing. Many provide camaraderie, skill development, and excellent recreation. Scuba diving, on the other hand, is considered by its enthusiasts as "far beyond" the realm of a sport or hobby. Diving allows one to visit and participate in a totally different world than most people ever even get a chance to see, first hand. Mankind lives on the *surface* of the earth, sometimes venturing onto the surface of the water. The world *under* water is seen by only a very few. It is as different as would be a visit to a different friendly and inhabited planet. The life-forms and landscape are vastly different, gravity seems almost non-existent, the "atmosphere" is over 600 times denser, and our sense of hearing is diminished almost to the point of un-importance.

And, above and beyond the wonder of the underwater world is the total sense of weightlessness experienced there. Suppose you could learn the game of golf well enough to get vast enjoyment from it. Great! You'll have a life-time sport that will provide hundreds of hours of valuable recreation. Now, suppose you could also enroll in a class which enabled you to achieve absolute zero gravity! You could learn to balance motionless on one little finger, hover effortlessly at any altitude you wished, move yourself easily a distance of 15 feet with only a few gentle strokes of your lower legs. In effect, *fly like a bird!* You might well still enjoy playing golf, when you're not diving, but your life would never be the same!

Scuba diving is, I am firmly convinced, vastly more than "just another hobby!" In the 40 years since I first learned to dive, I've had the pleasure of introducing hundreds of others to the experience. Countless times, former students have told me, years after their initial training, just how much diving has changed their lives, and added to their zest for living. Some also snow ski, as well. Or jog, play tennis, boat, make music, golf, or have many other interests and hobbies. But they tell me, over and over, that it is the ability to visit and experience the underwater world that has most enhanced the quality of their life. How rewarding, for me, to have had a hand in familiarizing them with the skills that initially permitted their participation!

Learning How to Teach

On the night we were first introduced to the use of scuba equipment in the indoor pool, during our own training, I promised myself that I would do whatever I could to make this same experience available to as many other people as possible.

This led me, after only our first couple of years of sport diving, to look into the possibility of eventually becoming an instructor. At that time, about 1960, no national agencies existed which were involved exclusively in scuba training or instructor certification. The YMCA Aquatic Program was just starting to get involved in the sport. On the west coast, the Los Angeles County Underwater Training Program was just starting to evolve into what eventually would become "NAUI" (the National Association of Underwater Instructors). The CNCA (Conference for National Co-operation in Aquatics), comprised of representatives of the YMCA, the Red Cross, and several other agencies, was just beginning to develop standards and textbooks. (The first handbook, The New Science of Skin and scuba Diving, with 11 chapters each prepared by a different author, was first printed in 1957.) Today's largest certification agency, PADI (Professional Association of Diving Instructors), was still several years from formation.

In Ohio, the diving clubs began to turn their attention from competitive events more to the standardization of diver training. Soon, several of these clubs joined to form the "Ohio Council of Skin and Scuba Divers" (OCSSD, an organization which still exists today).

Gradually, basic training standards were formulated, as well as a means of certifying instructors. Several dedicated people spent a good deal of time and study in the development of these programs, much of which was being learned in "pioneer" programs on the west coast. Eventually, the (Ohio) "council" felt ready to start training, testing, and certifying scuba instructors. The first such "Certification Institute" was scheduled to be held at the Toledo YMCA, the same place we had taken our scuba course. The whole affair was scheduled on a weekend, and intended to last about a day and a half. The cost, if I recall, was about $15. Though only members of clubs which belonged to the council were to be eligible, I was able to sign in as an "indeper[illegible] now, I had logged only about 35 hours of div[illegible] had read all of the books on diving that I could get my hands on. The only scuba class I'd ever had any experience with was the one we had initially taken to get certified. In looking over the council's printed materials and criteria, though, I felt that I might have a chance at meeting the mark. At my job, I'd had some experience in dealing with groups of people, and had taken some basic training in public speaking. One of my other hobbies, at the time, was stock car racing, and I was currently serving as the president of a local racing association, with experience at dealing with standards, rules, and organization.

Let me explain something here that non-divers may not be familiar with. Today, to participate in the sport, one must have proof that one has completed a course of adequate training. Clearly, diving is an activity that does present some hazards to the un-trained. The "proof of training" is in the CERTIFICATION CARD ("C-card") that one earns at the completion of a standardized registered course taught by a certified instructor. To earn this card, one needs not only complete all of the practical and theoretical work and testing involved in the course itself, but then also complete a number of supervised open-water dive experiences. Without the C-card, most dive shops will not refill an individual's air cylinder, clubs won't allow un-certified divers to participate in club activities, and charter boats and dive resorts will simply not allow the un-trained person to dive. This is an excellent practice.

In the "early days," of course, there was no such structure to the sport. Sport divers often just "showed" other people "how to dive." (Like our friend, Ron Heredon, had learned.) Even the class we took, at the YMCA, was not officially sanctioned yet, though we felt our instructor was quite thorough. "Wildcat" instructors were beginning to offer courses all around the country, but many of these classes were minimal, at best. Today, the national training agencies have evolved excellent standards and training techniques, and the modern

C-card from a recognized agency is accepted around the world as proof of most adequate competency.

I enrolled in, and completed, the Instructor Certification Institute at Toledo. There were only eight candidates, and only two of us passed. Some of the others, I learned later, followed up with more study and experience, and later earned their certification. We were tested in our knowledge of the physics and physiology pertaining to diving, and in our ability to teach. A brief pool test of our basic water and rescue skills was involved, but no work in open water. Years later, when I myself became involved with the testing and certifying of instructors, I could see that this first "Institute" was, indeed, quite primitive. (Today's new scuba instructor has invested many weeks of study, tests, and advanced skill perfection, and hundreds of dollars to earn certification!) However, it all had to start someplace, and here I was, in 1961, with an instructor's card and jacket patch!

A Teacher With No Students

Except for the card and the patch, however, nothing really changed. Our home town was too small to recruit a class to teach, and had no indoor pool in which to teach. My dive buddies probably benefitted somewhat from some of the "semi-advanced" skills that I learned at the Certification Institute and passed on to them, but otherwise we simply continued to sport dive whenever we could. We learned something from every dive....

One lesson we learned the "lucky" way. In Ohio, sport diving pretty well came to the end of the season in about early October, when the local lakes and quarries simply got too cold for our wet suits. Then, of course, by January, most were frozen over. Diving usually didn't start again until about mid-April. In the middle of the

This guy was determined to test the ice. Luckily, he had a dry suit on!

winter of 1962, Bob and I thought we might try an ice dive. We had heard that it was sometimes done, that the water tended to be unusually clear, once it had a layer of ice on top, and that the water, under the ice, was no colder than it was during the summer, in the cold layer under the thermocline. That is, about 54°F. And, that's ALL we had heard about ice diving. It was enough to cause us to want to give it a try. On a sunny Sunday, we proceeded to the nearest flooded limestone quarry, near Whitehouse, Ohio. With spud bars, axes, and hatchets, we cut a large hole in the ten-inch thick ice, over an area we knew to be about 40 feet deep. Indeed, once we removed the ice from the hole, the water was so unusually clear we could easily see the bottom! We went back to our car and suited up. However, by now the sky was clouded over, and a cold wind had developed. By the time we got back to the hole we'd cut, we were already getting chilled. Then, Bob dipped his thermometer into the water, and we got a reading of 32°! That clinched it for us. What little enthusiasm we had for the undertaking drained from us completely. We went back to the car, our wet suit tails between our legs, and dressed.

There were a number of things we *didn't* know about ice diving. First, the water in the ice hole itself, where Bob had dipped his thermometer, we later learned could quite well be expected to be almost at the freezing point. A foot or two deeper, and the water temperature was very likely to be that 54° that we expected. Later dives in other winters proved that to us. But, that day, that little thermometer kept us out of the water. Now, we know, that was the very best thing that could have happened to us! What we didn't know about ice diving is that, once under water, no matter how clear the water, the hole in the ice *absolutely disappears*, once one has moved only a few feet away from it! One of the cardinal rules of ice diving (now we know), is to NEVER dive without a life-line attached to each diver and to an immovable object topside, so that one can always find the way back to the hole! We simply hadn't heard about that neat little life-saving practice, and if we had entered the water that day we may well have not been found till spring!

This reminds me of another ice-diving adventure in later years, in which I was not a direct participant. A club in central Ohio often dove in flooded strip mines, in an area formerly mined for coal. Some of these lakes were remarkably clear. One February day several members of the club gathered for an ice dive at one of the lakes they were not yet entirely familiar with. Armed with chain saws, they quickly had the recommended 6-foot by 6-foot hole cut in the thick ice. As per suggested ice-diving practices, they shoved the intact cut slab back under the ice (so that it could be re-floated when they were done, quickly freeze into place, and prevent a hazard to ice skaters, etc.) Once suited up, with all of the appropriate safety ropes attached, the first of the divers prepared to make his entry. A photographer stood by to record the moment. Grasping his mask in the classic manner, the diver executed a perfect feet first entry (a simple "giant step" into the water.) Perfect! Only one small problem existed. Unknown to the club, they had cut this perfect ice-diving hole in an area of the lake that was only two feet deep! The first diver, performing his picture-perfect entry and expecting a complete submersion, was, instead, now standing in water only thigh deep! Extending his hands in a palms up "what now?" position, the diver was captured for posterity on the photographer's film. When the picture was later circulated, the diver tried to explain that it was simply a "stop action" photo that caught him in mid-entry, but, of course, the careful observer could see that the water in the hole surrounding his legs was now perfectly still.

The Underwater Angler

Another adventure took place during a dive in an Indiana lake. Like many lakes in the region, it was thought to be "fished out." Divers knew different. This one was loaded with monster-sized small mouth bass. During the summer months, however, these bass were all to be found in deep weedy ledges, mostly abut 30 feet deep, and unknown to most fishermen. And, apparently, they had enough natural food so as not to be often enticed by fisherman's baits. A funny thing occurs when divers enter an area where bass abide. The bass, though lethargic before, get excited and feisty when divers are in the area. (To fish, divers and their bubbles are tremendously noisy and obtrusive, and bass, particularly, get highly agitated!) Often, just swimming past such a submerged weed patch would cause the bass to dart out and seem almost aggressive. This day, early in the dive, Bob found a fisherman's lost artificial nightcrawler lure, with the swivel and about 20 feet of monofilament line still attached. Inspired to "troll" this lure past a weed bed, Bob wrapped several loops of the line around his hand, and began trailing the lure about 15 feet behind himself.

In short order, a "grandpa" smallmouth streaked out of the weeds, swallowed the lure, and, feeling the hooks, headed like a scalded cat for deep water. Exactly what Bob had hoped for had happened! Except there was no stopping this fish, and Bob had no reel to bring him in. Instead, the fish line wrapped around Bob's hand and fingers laid some deep and painful cuts in his hide! This bass wasn't gonna' stop! By the time Bob cut the line with his knife, he knew he'd have some lasting scars with which to remember the encounter. In fact, our diving for the day was done by the time we got him doctored up. I suggested that it might be safer to use his blowgun on future bass.

Just to further illustrate this agitated bass concept, I'll describe another incident that occurred in Lake Erie. We set out to dive the wreck of the *Adventure*, a wooden schooner sunk just off the east side of Kelley's Island. The wreck was in only about 25 feet of water, and was a favorite fishing spot. Sure enough, when we got to the area, two boatloads of fishermen were anchored to the wreck. We asked both boats if they would mind if we dove the wreck. "No," they said, "go right ahead. We haven't caught a thing here all morning, and we were just thinking about leaving." When we reached the bottom we saw that the wreckage was loaded with bass, as usual. And, as usual, as we swam by their lairs, we could see that they became highly excited. We proceeded with the dive, collected some of the china and brass "artifacts" the wreck usually produced, and surfaced about 45 minutes later. The fishermen were still there, and were obviously more excited than the bass! They told us that almost immediately after we submerged, the bass had started to bite like a herd of hungry hogs. In fact, in that 45 minutes, almost every one of the fisherman already had his limit! I've never heard a scientific explanation of this kind of fish behavior, and it seems most pronounced with freshwater bass, but it is definitely a phenomenon that has often been documented!

Turtle Attack

Another incident involved Bob, Dick, and I, and a dive in a rather large lake in southern Michigan. Our employer's sales staff had a weekend poker and fishing party at one of the guy's cottages, and we three took along our dive gear, in case we'd have a chance to use it. Early the morning of the second day, we submerged from the dock on a compass course straight out into the lake. Gradually, as we progressed, the depth increased to nearly 50 feet. The bottom, however, was nothing but featureless sand and mud. Looking closer, we began to spot some sizeable turtles partially buried in the bottom, almost as if they were semi-hibernating. We found that if we snuck up on them from the rear, and poked their shell, they would take off like a rocket, quickly disappearing into the distance. Cal spotted a really big snapper, probably a 20 pounder. He headed toward it to see if the same reaction would occur. Dick was kind of hanging in mid-water, about 15 feet over Bob's head, casually observing this new "game." As it happened, Dick's weight belt had about a foot of extra length, and this surplus was simply hanging down from under him. Bob poked the big turtle. It exploded from the bottom, as expected. Instead of streaking out of sight, however, it cut a straight line for Dick's mid-section. Before he could even react, the turtle took hold of the loose end of Dick's weight belt, and started tugging, like it wanted to take Dick home to show his kids. Dick, completely startled, could only react by trying to back away from the turtle. After all, he said later, who knew when the beast might try to get a better grip on some other nearby appendage? Bob, thinking that if this dumb reptile might take off with Dick, or some part of him, that he might be held responsible, immediately attacked the turtle, beating on his shell with the butt of his knife.

This whole scene went on for what seemed like several minutes, until the turtle eventually released his grip on Dick's belt and took off into the blue. We could only guess what might have motivated him to behave this way. He may not have seen Bob, the guy who thumped him, and presumed that innocent Dick was the guy who disturbed his rest. As to what he thought the loose weight belt was — who knows? We didn't do any more turtle thumping, in any case, and I don't think any of the guys back at the party ever did believe our snapper story, even though you could clearly see the turtle's mouth print on Dick's weight belt!

5

Old Dogs Learn New (Underwater) Tricks

In 1960, my employer, a meat packing company, promoted me to the position of regional Sales Manager, and moved me and my young family to Newark, a somewhat larger town in east central Ohio. They had bought out a smaller competitor there, and I was assigned the job of "sorting out" and re-orienting the existing sales staff.

Unlike my home town, Newark had a YMCA with an indoor pool. Furthermore, as I soon discovered, a scuba club existed there! The Air Force had a large facility in Newark, involved in the calibrating of missile guidance systems. About half of the 400+ employees of the facility were military, the rest civilian. The core of the local diving club consisted of some of the employees who had been transferred in from a similar facility on the west coast, where they had been exposed to the new sport of scuba diving. Others were local sportsmen who had simply become intrigued by the concept of diving, much the same way as I had.

I was invited to join the club, the Newark YMCA Poseidons, and learned that none of the 12 members had been "officially" trained and certified. While they actively scheduled dives in local lakes, quarries, and rivers, and had an excellent safety record, all had received their training by "informal" means. Upon learning that I was a certified instructor, they quickly proposed that I organize a class through which they could all earn authorized credentials. Since the club was affiliated with the YMCA, and met there, the pool was available for this training. The Aquatic Director at the YMCA was enthusiastic about the project, as the National YMCA program was just getting into scuba training as a new feature of their excellent aquatic program. However, in order to teach a "sanctioned" YMCA course within the program, it would be necessary for me to achieve YMCA instructor certification, even though I already had been certified through the Ohio Council.

By now, I'd been diving with the club members several times, and could see that they were quite well skilled in diving and most of the accepted safety practices. So, we had a situation where I was now expected to "certify" these guys, most of which were already quite as experienced as I was. And, to do so, I would have to "step up" to what was almost certainly going to be a higher level of testing, to earn the certification I needed to certify them. Remembering the promise I had made to myself to "pass it on," I determined to proceed. First, I warned the members that, once I earned YMCA certification and started their class, my principles would not allow me to simply certify them on the basis of their evidenced experience. Once I learned exactly what the YMCA standards were, I would feel obligated to put them through all of the paces just as if they were beginners, even if this proved embarrassing to me or to them. They unanimously agreed with the concept, and urged me to proceed.

To earn the YMCA certification, I would need to attend a two-day institute. One of the first ones ever to occur in Ohio was scheduled for about six weeks hence, in Dayton. I sent for and received the registration materials. As expected, this was going to be a lot tougher than was that Ohio Council affair in Toledo! First, I came to realize that the YMCA expects all of their Aquatic Program Instructors to be extremely proficient in ALL water skills, including life-saving, and also be quite physically fit! By now, a YMCA scuba textbook also existed, and it involved a great deal of physics and physiology, which I would be expected to master. I had no choice but to hit the pool, study the book, and exercise my way to a better level of fitness. The Certification Institute was to start with a "basic" swimming proficiency test which involved more surface and underwater swimming and water work than I had ever done. The aquatic director at the Newark "Y," Don Watson, worked almost every evening with me, to help me prepare, and in the process also making me familiar with the YMCA philosophies and methods. (He later enrolled in one of my courses, and became a most enthusiastic diver!) Luckily, I was also able to enroll in, and almost complete, a full lifesaving course.

Off to Dayton, finally, but still with a great deal of apprehension. The staff at the institute was comprised of certified instructors from neighboring states, YMCA aquatic directors from the local regions, and a representative from the National YMCA headquarters. I joined with some 14 or 15 fellow candidates, and soon realized that several were actually far less prepared than myself. (In those days, the cost for these Certification Institutes was so slight that a goodly number of sport divers signed in, just "to see if they could do it," and/or to get another patch for their jacket, without really having any intention to actually teach the sport.)

At this time, in the early stages of making the training more readily available to the general public, several of those who were in charge of setting the standards had been exposed to the military methods used in training Navy "Seals," "Frogmen," and Underwater Demolition Teams (UDT). Those already certified as instructors, and now being in the position to certify another generation of instructors, were determined not to "lower the standards." If a part of their own prerequisites was to be able to run five miles to get *their* instructor certification, then, by God, anyone else who was going to get that same certification was going to have to run at least five miles — and maybe a few more, just for good measure! If they had been grilled and "nailed to the wall" during *their* oral presentation evaluations, then any candidates who faced them were going to be in for a real "third degree." If *their* written examinations had been tough, then, by golly, they would add some questions to the exams they developed which would make a physics professor cringe!

In later years, I was to become deeply involved in the operation and evolvement of the YMCA scuba program, serving several years as the (volunteer) regional director for a three-state area. Gradually, we were able to work some of these macho personality differences out of the instructor certification process, and standardize the tests, procedures, and requirements, etc. In the process, the YMCA scuba textbook was revised several times, and became more oriented to civilian sport diving, as opposed to military standards.

The Certification Institute at Dayton, however, proved to be an experience in encouragement. While some of those on the faculty staff were "Marine"-minded, the Director, Bob Farmer, fully realized that the candidates would benefit from some training and orientation, not just testing procedures. After all, at that time there were no pre-certification training seminars, nor any means of preparation except for one's own individual initiative. (Later, we would evolve a series of teaching sessions, to precede participation in a Certification Institute.) Bob, at the Dayton Institute, had briefed his staff to incorporate all of the "teaching" possible, in the time available between the testing sessions. This proved very helpful, though the time to do so was substantially limited.

Still, Farmer couldn't be everywhere at once, and some of the faculty remained determined to lay some booby traps for the candidates. A couple of the candidates resented this approach, and simply walked away. A few others got into what could mildly be called "personality conflicts" with certain of the trainers, and allowed their exasperation to negatively affect their performance. Most, however, accepted the harassment good-naturedly, and, like myself, determined to get all the benefit they could from the exercise, pass or fail.

The most difficult portion of the process was the evaluation of the candidate's oral teaching skills. We had been directed to prepare a lecture on a topic of our choice, and present it in five minutes. Two other evaluation sessions were "extemporaneous," meaning that we were called before the evaluators and given a topic to "teach." Now, it is understandable that this process would be stressful to the candidates. A series of three rooms were set up, with a different set of three evaluators in each room. We were directed to make our presentations "just as we would to a group of beginning students." Sure! The volunteer evaluators, of course, were not professional educators. In fact, most were "blue collar" guys, who couldn't help but remember the hell they had been through, when they were on the other side of the table. Some tended to have some empathy for the situation, while others seemed determined to trip us up with tricky "student type" questions and responses. As to the prepared five minute lecture, some of the evaluators seemed to expect us to thoroughly cover and summarize the entire topic in that five minutes (virtually impossible), while others indicated that they simply wanted to hear a five-minute segment of a prepared presentation, whether there was time to conclude it or not.

At the conclusion of the institute, as the results were announced, it developed that two of the candidates who had failed were professional school teachers. They deeply resented that their teaching abilities had been found lacking by judges who were not themselves educators. The staff answered that while the individual's teaching skills and manner did, indeed, seem clearly professional, they had evidenced an inadequate knowledge of the subjects (as was also evident from their test scores). Both stormed from the room, despite the staff's urging that they gain more experience and knowledge, and try again at a later date. In later years, I would witness this same scenario a number of times, and became convinced that some teachers seem to feel that if they know how to teach, they can teach any subject adequately well. I don't happen to agree with that concept, though a good teacher who does master the diving topics (and who also gains a wide variety of diving experience), can truly make an excellent scuba instructor!

Three candidates did pass the Certification Institute, and were immediately awarded their credentials. Four others, including myself, passed "provisionally," with certification to be awarded when they had completed a few of the prerequisites they were still missing. In my case, all that was required was that I complete the Lifesaving course back at my home YMCA, which I did within the next couple of weeks. Incidentally, I became good friends with several of my fellow candidates, and some of the staffers, and have enjoyed many rewarding diving adventures with several of them through the years.

Now I was a full-fledged CERTIFIED SCUBA INSTRUCTOR, and qualified to set up and teach sport divers, and issue official YMCA (and Ohio Council), training credentials!

As became apparent when I had started diving with my new friends in the Newark club, it is rather incumbent on someone with an instructor rating to conduct oneself in as professional a manner as possible. In other words, I could sense that I wasn't expected to make any errors in my diving procedures. I tried my best to avoid any "boo-boos," while at the same time trying hard not to appear that I thought that I was any more "professional" than the others. (I wasn't, really!) I could also sense that some of the members were somewhat alert to catch me up, if I did "slip up." Now and then, of course, I'd be caught in a minor goof, but I was always careful to laugh at myself, and I eventually felt that the "pedestal" factor was of less concern than I feared.

"Murphy's Law," of course, dictates that when an instructor does goof, a student will be watching. An example of this occurred to a west coast instructor, with whom I later became acquainted. He had a new class of eight students on a boat, for their first ocean dive, just off Catalina Island. He assembled the students on the stern of the boat, and directed them to all sit on the railing, as he was. Step by step, he instructed them to don each piece of equipment, in the proper order, first demonstrating by donning his own. Soon, they were all equipped. The instructor described the "back roll" entry they were all to make, that is to simply roll backwards off of the rail into the water, about six feet below. They were to do this in order, following his own entry, then once in the water they were to all rendezvous on the surface, at the anchor line. He directed them all to make a final check of their equipment, which he also performed on himself.

While the students were busy looking over their own equipment, the instructor noticed that he himself had forgotten to secure the "crotch strap" that passed from the rear neck of his inflatable life preserver, through his legs, to a "D" ring on the bottom of the front of the vest. Since no students were watching him at that moment, he simply reached down between his legs, pulled the strap up, and clipped it to the "D" ring. Great! No one had even noticed! However, the strap had been hanging *outside* of the rail on which he was seated, and it was now, unknown to him, cinched *around* the railing! He signaled his students that he was about to make his entry, and reminded them of the order in which they were to follow his lead. He then placed his regulator mouthpiece in his mouth, grasped his mask, and performed a picture-perfect back-roll. Of course, the crotch strap proved strong enough to hold his weight, and caught him in mid roll. There he hung on the railing, fins in the air, going no further. The students gaped. There was nothing for the instructor to do but to reach for his knife and cut the strap, dropping ignominiously into the water. Embarrassing! Later, the students told him they held a brief conference before they made their entries, wondering if they, too, should re-position their own crotch straps, and do it the way the instructor had "demonstrated." (They didn't.)

Instructors are human, of course, and I, too, have had my full share of "humbling experiences." It's true enough that sport diving is never quite the same for the diver who gains instructor certification. For one thing, an active instructor usually finds that most of his time spent at dive sites is involved with the training or supervision of new diving students. In season, little time is left for the instructor to participate in diving just for the joy of it. It is somewhat a case of a hobby becoming a vocation, of sorts, with less time being available for informally enjoying the hobby. Nonetheless, the trade-off is well worth the reward of being able to introduce numbers of other people to the safe conduct of a sport that will influence the rest of their lives. The instructors that I have known who I would rate as among the best are those who always found great enthusiasm in the sharing of their knowledge. Those who instruct for the "prestige" or the money are much more likely to be mediocre and un-inspiring.

So, here I was, in 1962, ready to put together my very first scuba course, with my first students to be a group of divers who were already diving! Even with the club members' enthusiasm and understanding of the situation, I knew that a great deal of "diplomacy" was going to be required. I developed my course outline, and distributed it to the students, in advance of the start of the course, so they would see what was going to be involved. I clearly identified those procedures and requirements which were dictated by the current YMCA standards, and outlined the process by which I intended to prepare them to meet these standards. I felt that this advance knowledge would help the "students" understand the training (and would also prevent anyone of accusing me of "making things up," as the course proceeded!).

As it developed, the most difficult part of the whole undertaking was the "physical fitness" standards which the YMCA then insisted on. Before proceeding with any of the skill training, it was necessary that every student pass a strenuous fitness test. This involved a series of water skill exercises nearly the same as I had had to complete prior to my YMCA instructor certification, and a series of gym exercises, such as push-ups, wind exercises, pull-ups, etc. It was at this point that a couple of the club members came to me privately and asked if they could be "excused" from this part of the requirements. This was understandable, since they were clearly out of shape. However, the philosophy at the National YMCA headquarters, at that time, was that a scuba diver might, at any time, be required to go to the aid of a diving buddy in trouble, and must be physically fit enough to render full assistance. This made sense, and still does, though with today's modern safety and rescue equipment, physical strength is not considered as important a requirement.

The two individuals who requested a personal "variation" from the physical fitness standards happened to be those who had most readily welcomed me into the club in the first place, and who had most enthusiastically promoted the concept of my teaching of the certification course for the club. Their plea was sincere, and, on the basis of their lengthy safe diving experience, realistic enough. However, I could not comply. At the risk of the loss of their friendship (and, perhaps, my future with the club), I *had* to find a way to get them through the test, with no exceptions. This, of course, put them on the spot, too, because my decision could easily cause them to "lose face" with their fellow club members. I had left them only the choices of dropping out of the class before it started, proceeding with the physical fitness tests and failing them, finding a way to talk me out of my "integrity," or, somehow, becoming capable of passing the tests. I pushed for the latter, of course, and offered to work with them privately on a program of preparation, which we did.

In the process, one of the individuals, who happened to be the one most influential in the club, and also substantially overweight, almost came to blows with me. One of the requirements involved running a mile, within a certain time limit. This fellow could not run even a quarter mile, at the start of our private sessions. Gradually, we built him up to a mile, but it was his absolute max, and he could not get his time down to the requirement. Each time he attempted it, and totally exhausted himself in the process, and I had to show him the stopwatch, he was highly distressed, to say the least! I was doing this whole thing on a volunteer basis, and giving extra of my private time to help these two guys prepare for it. Not much gratitude was evident, however, and this one fellow, particularly, grew quite impatient with my insistence. (He used several other words for it...) Just as I was beginning to fear for the whole project, however, they both finally were able to pass the test. Later, they would laugh about their difficulty, and both privately indicated that they felt a real sense of pride in their accomplishment, but it was really touch and go for a while!

Our course was scheduled to meet one night each week, for a total of seven weeks. At that time, the YMCA standards did not require any field trip (open water), dives for certification, but I added two local dives at the end of my course, and the club members agreed with the concept. They had little trouble with the text book work and tests, and mastered the required water skills quite nicely. Since the local YMCA still owned no scuba equipment with which to teach, these students simply used that which they already owned. As the course progressed, however, many of the club members related that they were picking up a good deal of valuable information and new skills. This was a good group! Even the two men who had so much difficulty with the controversial physical fitness did quite well (and the big guy had lost some weight in the process!). During the classroom sessions, I had feared that some of the guys would be "impatient," if not bored, but all seemed pleased that they were picking up new and useful information. In any case, they made it seem far less difficult for me than I had expected.

In the end, every one of the club members earned their certification, and they all appreciated, as I did, that absolutely no exceptions had been made to any of the required standards! And, especially counting the supervised open-water work, they received a great deal more training than Bob, Dick, and I had obtained from our initial course. I was rather pleased about that, and the members couldn't have been prouder of their new cards and patches!

Just Add Water!

Don Watson, the Newark YMCA aquatic director, was eager to set up more scuba courses, and we immediately scheduled a number of them. This was made somewhat difficult by my work schedule, which involved trips out of town on most week days, but I was able to work around this. All of the newly-certified club members offered to assist with these classes, and to loan their personal gear to teach with. We decided to accept classes of up to as many as fifteen students, and soon had no trouble in filling three or four classes each year. The majority of the graduates of these classes joined the club, which soon grew to one of the largest in the State.

My eager assistants and I volunteered our time for these classes, for the first few years, and the students were required to pay only for their textbook, credentials, and a modest fee to the YMCA. The course proved so popular that the YMCA eventually purchased several sets of equipment to teach with.

At this point, the National YMCA standards still did not yet require the students to participate in any supervised field trip dives prior to certification. However, instructors were permitted to ADD any requirements to the basic standards that they deemed appropriate. I personally felt strongly that open water work should be a prerequisite, and required my graduates to participate in at least two structured dives, usually in local quarries or lakes. In each case, the new student would be accompanied by an experienced dive buddy, either myself, or one of the club members. While a few skill practices were required, such as mask clearing, these "check-out" dives were mostly intended to be pleasant introductory experiences.

At this same time, instructors on the west coast were also experimenting with the incorporation of open-water dives for their students. These exercises, however, were often designed to be severe tests of the students' reactions to simulated "emergency situations," and sometimes often involved strenuous and/or extremely uncomfortable exercises. (The "UDT factor" was still prevalent, with many instructors.) Many students were walking away from the sport, after these "military" experiences, having experienced such stressful circumstances on their initial few dives.

At about this same time, the newly-formed NAUI (National Association of Underwater Instructors), launched the concept of annual "conventions" of U/W Instructors, to exchange ideas and further advance the sport. The first such meeting was held in Anaheim, California. I had become affiliated with NAUI, through my work with the YMCA (I became NAUI INSTRUCTOR A-40), and was accepted to present a paper at this convention. The keynote speaker was Jacques Cousteau, who was also in the audience for my presentation. (Talk about butterflies in the belly!) My topic was the value of making the field trip dives a required part of the student's training, and making these initial dives mostly pleasurable experiences, versus the "dives from hell" that some instructors seemed to insist on. Naturally, the military-type instructors in the audience felt that I was proposing a "lessening of the standards," which they opposed. ("Hey, by God, I had to do this or that, and if it was good for me, it will be good for my students! I want *my* students to be TOUGH!")

That evening, at a cocktail reception, I had the only conversation I've ever had with Commander Cousteau. It was quite brief. He said to me "You spoke on a very controversial topic today." "Yes indeed," said I, thinking I was about to partake in a scholarly conversation with "Mr. Scuba" himself. "I am inclined to agree with your viewpoint," he said. End of discussion. He then drifted off among the crowd. I was encouraged, however, and continued to press the issue, through the years, at every opportunity. Eventually, as others of a like mind helped with the process, open-water dives did become a certification requirement with all of the training agencies, and most instructors started to provide more comfortable conditions for their students' first few dives.

I, too, certainly believe in training divers who are well-prepared and capable. In fact, 30 years later, I still insisted on providing training well above the "standards" to my own students. I did this, however, by providing more thorough and lengthy training, and more open water dives, instead of putting them through underwater endurance tests. This became more and more realistic and practical, as "advanced courses" were designed and made available to those students who wanted to progress to specialties such as "search and rescue," "cavern diving," etc. They could either stop with their initial sport diver training, or proceed on with continued classes which involved more detailed skills.

In the meantime, our local club continued to grow and become ever more active. Trips were scheduled to the Florida Keys, Canada, the Bahamas, and other Great Lakes. A number of the club members' wives, and several other females, began to enroll in our classes. The club became closely associated with the Ohio Council, and many new friendships evolved. At the same time, the growing council began to stage numerous events which brought the clubs together. Outdoor summer competitions were revived, and winter indoor "U/W Olympics" developed. An annual banquet was initiated, providing opportunities for inter-club U/W photo contests, "artifact" displays, and, eventually, well-known guest speakers. Our Newark Club won many of the competitions, and several of our members, including myself, served a number of posts in the council, which became one of the most respected in the nation.

It became an annual event for the club to visit the Florida Keys each August, at the start of the lobster season. Most of these trips were family camping affairs, though we also sometimes found accommodations at "economy" motels. At this time, the business people in the Keys were only just starting to become aware that scuba divers might actually represent some income for them. Few boats were available for diving. Usually, we had to find a fishing charter boat that didn't happen to have any fishermen booked. Only about three air refill stations existed in all of the Keys. (Today, there are over 70, and dozens of dedicated diving charter boats!)

While nearly everyone in the club liked fresh lobsters, a couple of the fellows went absolutely lobster-crazy. One diver in particular, Sam Berkely, couldn't get enough. Sam had spent some time diving on the west coast, where diving is more game-oriented. There, he had a reputation as being the "abalone king," and also liked scallops so much that he would sometimes shuck and eat them under water! Sam was a good-looking guy, probably around 25 at the time, and a fine physical specimen. He was also a guy who didn't care that much about rules and regulations. One year, when lobsters were rather scarce on the reefs, he decided to catch some at night, in the boat harbor at the campground. Now, it was illegal to lobster at night, and illegal to lobster within a certain distance of the overseas highway. The boat harbor was within this "no lobstering" zone. Sam couldn't get anyone else to accompany him on this felonious dive, so he went alone. (A cardinal sin, in safe diving.) He surfaced, in short order, with a burlap bag almost full of lobster tails. Since most of the lobsters in the harbor were under-sized, he simply broke their tails off under water. (Also illegal.) He then invited everyone in camp to join him in a big lobster boil. Everyone declined, not wanting to be a part of this whole thing. He and his wife had a lot of lobster to eat by themselves. While several of us reprimanded him, nobody "squealed," and he wasn't caught. (We calculated that, counting all of the infractions he could have been fined for, he could have been nailed for some $50,000! That would have been an expensive meal!) He either lost some of his taste for lobster or got our message, and didn't dive the harbor again!

Florida lobsters, by the way (Spiny Lobsters), don't have claws, like Maine lobsters. The only edible part is usually the tail. Still they are elusive, and not easy to harvest. Being also covered with sharp spines, they can be a bit "ouchy" to handle, and, in Florida, they can only be caught by hand, by divers. A new diver usually asks, "Exactly how do you go about grabbing a live lobster?" The answer, basically, is "Any which way you can!" Someone once answered this question by saying "Suppose you were on the deck of a boat on a windy day, and a fifty dollar bill blew in and landed on the deck, and was about to be blown off the other side. Exactly how would you grab that fifty dollar bill? That's *exactly* how you would grab a lobster!" In other words, go for it first, and worry about your technique later!

Eventually, I became rather skilled at harvesting lobsters. (The more you like to eat lobsters, the better you tend to be at catching them!) Another year, I escorted a group of recently certified divers from a local university to the Keys, for a lobstering expedition. All were anxious to catch and devour as many lobsters as possible, and on the trip down they had hundreds of questions about exactly where we would find them, and exactly how they should go about catching them. A more eager group of neophyte lobsterers couldn't exist. For our first diving day, I had chartered a boat with a skipper I had previously been diving with, and who knew how to find the best lobster grounds. A new limit of 24 lobsters per boat had just gone into effect, so the message for the day was that we would come back with the 24 BIGGEST lobsters we could find, even if that meant replacing some lobsters already on board with larger ones, as they were caught. This group was ready! They had been talking lobster for 1,500 miles. They were in the water in short order, and it proved to be a great lobster area. I have never seen more lobsters attacked in one dive. Despite their enthusiasm, however, or, perhaps, because of it, they couldn't manage to get the "bugs" into their bags, try as they may. By the end of the day, however, we had 24 lobsters on board, some of which were second or third generation, size-wise. I had personally caught 23 of them. Among these 12 young eager lobster grabbers, they had only managed to get one in the bag! We had a great lobster feast that night, with sweet corn and salad, and, by the end of the trip, most of them were doing far better at "bugging."

This reminds me of another instance that involved Maine lobsters, the ones with claws. Two fellows from the Boston area associated with the diving business, and both named Frank, went for a lobster dive off the Maine coast. One of these guys is only about five feet tall. I'll call him Short Frank. The other Frank is well over six feet tall, is an excellent diver, but has only one arm and one leg. I'll call him Tall Frank. Short Frank was an accomplished lobster grabber. Tall Frank had never been lobster diving. Short Frank was going to show him how it is done, and was determined to start out with as large a specimen as he could find. He soon located a whopper of around six pounds. (Lobsters this size have big claws!), and motioned Tall Frank to come closer to observe exactly how he was going to capture this beast. With Tall Frank watching, Short Frank positioned himself just right, and made his famous lightning-fast GRAB. As Murphy's Law would dictate, Short Frank's grab didn't succeed, and the lobster retreated further into its hole. Short Frank, with his short arms, now couldn't reach it. This was such a nice large specimen, however, that Short Frank didn't want to abandon it. He motioned Tall Frank, who only had one arm, but a far longer arm than Short Frank, to reach back in that hole and grab that lobster, something that Tall Frank had never done. He was game, though, and did as Short Frank had indicated. As he reached deep into the hole, he was mask-to-mask with Short Frank. Short Frank describes the scene. "I was looking into (Tall) Frank's eyes, as he reached back into that hole. Suddenly, his eyes got bigger, and tears began to roll from them, right inside his mask. I knew at once that Frank didn't have that lobster, the lobster had Frank!" Short Frank later said that he would never have believed that such an exquisite expression of pain could be read in one's eyes alone. In the end, however, even though his one hand now carries some lobster scars, Tall Frank *did* bring that lobster out of the hole, and thusly, I gather, passed his lobster initiation rites with Short Frank!

In southeastern Ohio, dozens of lakes are located in a vast area once strip-mined by the local power company. The whole area is now a designated wilderness, and the lakes were created to somewhat enhance the total environmental destruction wrought by the early methods of surface coal mining. Still owned by the power company, the area is administered by the Ohio Division of Wildlife. In theory, it is called a recreational area, though the only roads that exist are old mining roads. Basically, it is used by hunters and a few fishermen. Some of the lakes sustain a population of game fish, while many others are so polluted with coal sulphur and other chemicals that few life forms can exist. Our scuba club, and another from Zanesville, scouted out the few of the lakes that were clear enough to dive. They soon learned that some of the waters were so chemically toxic that the chrome on their diving equipment turned black.

Many of these lakes were wooded valleys when they were flooded, and the trees were left standing, to eventually rot away. It proved to make for fascinating dives, to "fly" among the branches of these still standing submerged trees. In fact, one could sometimes spot intact bird's nests, in some of the trees, and, believe it or not, we sometimes even found eggs in some of these nests! The bottoms of some of these lakes were comprised of a layer of semi-suspended silt, sometimes several feet thick. In the clearer such lakes, the top of this silt layer appeared to the eye to be the actual bottom. However, the diver could swim straight down into this, and disappear, just as though he had penetrated the actual lake bottom. We filmed some movies of this, and the image of the diver emerging back up out of the "bottom" was quite spectacular, as he trailed a giant cloud of silt behind him.

While diving in one of the better lakes one day, we took a break for lunch, leaving our equipment on the weedy shore. About ten minutes into the afternoon dive, at a depth of around 30 feet, my buddy suddenly gripped my arm and brought his face in front of mine. He eyes were as big as saucers, and were rolling wildly, and I at first feared that he was suffering some new exotic diving malady, or having some sort of fit. Then, as I looked closer, I saw that a bumble bee was flying around inside his mask, and his eyes were simply following the flight of the bee. Apparently, while his mask was laying on shore, this bee had taken up residence in a corner of the mask, staying put and unseen until we were well into the dive. This was at once both comical and startling. After all, this agitated bee could easily sting my buddy in the eye! Terrified of this possibility, he didn't immediately know what to do about the situation. The only action I could think of was to grab his mask and rip it from his face. The bee, of course, was ejected into the water. We joked later about whether or not the bee ever made it alive to the surface, and whether he suffered the "bee bends" later, if he did. For those few moments under water, though, as we both watched the bee zinging around inside his mask, it was rather an unforgettable scenario!

During one of our club's "economy" trips to the Virgin Islands, one of our newer members, Floyd McKenna, was observed collecting small hand-sized sea fans. Back then, the taking of coral specimens seemed not to be thought of as being truly destructive to the reefs. We were curious, though, why Floyd wanted so many of this one type. Floyd, frankly, was a character who seemed to view life somewhat through a happy fog. Never a drinker, he nevertheless usually behaved as though he was in the early mellow stages of a happy drunk. Back home, he operated kind of a "free spirit" shoe store. A friendlier, more pleasant guy couldn't be found. On this trip, it occurred to him that these little colorful sea fans would make great gifts for the folks at home. He also thought they would make unusual "tips" to hand out. Once we realized his plans, we cautioned him that all such specimens taken from the sea had to be carefully treated with preservatives, then thoroughly dried. Otherwise, they STINK! Heedless, he simply stacked up the sea fans at the end of each day, and, when we packed for our return flights, he just stuffed his several dozen specimens in his duffel bag and carry-on.

By the time we cleared customs, Floyd's bags smelled so bad you could almost see fumes rising from them! The customs agents wouldn't have anything to do with them, sending them on down the line un-inspected, as quickly as they could. On our flights, Floyd's carry-on bag by itself stunk up the passenger compartment so badly the attendants used up several cans of deodorant spray. Like when someone silently breaks wind in church, however, the culprit couldn't be identified. That is, until near the end of the flight, when Floyd decided to retrieve his carry-on bag, open it, and hand out these absolutely rank sea fans to all the attendants, grinning from ear to ear as he made this marvelous gesture. ("Anybody can give tips of money, but this will be such a unique gift they will treasure and keep it forever!") Clearly, Floyd had no sense of smell! Each recipient smiled weakly, and rushed to the galley or bathroom with their new "treasure " and, I'm sure, quickly disposed of it. Floyd was beaming, and he repeated the whole process on the next leg of our flight, and in all of the restaurants and coffee shops in between, much to the consternation of everyone present. Despite our continued pleas of "Ditch 'em, Floyd!" he was sublimely oblivious to the cloud of almost poisonous odor he unleashed at every opportunity. What a trip!

For another club trip, we booked economy flights and accommodations to dive Little Cayman Island, south of Cuba. The real attraction here was an area just offshore of a neighboring island, Cayman Brac. A cove here is known as Bloody Bay, and features a vertical drop-off starting at a depth of only about 30 feet. Many other tropical islands are also surrounded by such vertical walls, but often the upper edge of the wall is so deep that one can only go "over the edge" for just a few more feet. "Wall diving" is extremely popular, especially because numbers of large pelagic (open-ocean) fish can often be seen there. While photographing coral, sponges, and reef fish, the diver often spots larger fish not usually seen near a reef, such as manta rays, eagle rays, sharks, jacks, and perhaps even tuna or marlin. The deeper the bottom of the wall, the more likely it is to see something unusual, as it passes by. The foot of the Bloody Bay wall extends to a depth of several thousand feet. Of course, we would limit our depth to only 60-80 feet, but even here one has a sense of being suspended in mid-water, knowing full well that if you "dropped," there is nothing below you but nothing. It is here where the sensation of "free flight" is truly savored!

To and from our rooms and the dock, on this trip, we passed a thatched native "bar," where several islanders seemed to maintain a perpetual poker game. They seemed friendly enough, when I'd stop to watch, and the stakes weren't too high. One day I asked if I could sit in on the game for a while. They readily made room for me and my money. By about the fourth hand, I found myself with two pairs, and when the bet came to me, I raised. None of the other players dropped out. When we showed our hands, I clearly had the best. I reached for the pot, and one of the gentlemen put his hand on mine. "That's MY pot, sir!," he said. I looked at his cards and he had no pair, no flush, no straight, no anything. I asked why he thought his hand beat mine. The other players went silent and their grins vanished. "Why sir," he said, "I have a Cayman Full House!" "Perhaps you don't know our rules," another player added, and pointed out that the "winner's" hand was comprised of three spades and two hearts. I was advised that this hand was better than even three of a kind. I'm sure that there are plenty of "old sayings" regarding strangers playing poker by somebody else's rules. In any case, there wasn't that much money involved, and all of my friends who had been watching the game had suddenly found they needed to be someplace else. I laughed and apologized for not having recognized this fine hand the gentleman had. I also happened to remember, just then, that I, too, needed to be elsewhere. The game resumed without me, and I've never learned if there really is such a thing as a "Cayman Full House," or if I was simply a fish. As I walked away, the players were chuckling, but I don't know if it was because of my lack of Cayman poker knowledge, or my lack of good sense.

One winter, in Newark, the diving club decided to stage an ice-diving "demonstration," both for the publicity and as a means of training newer members in this specialized off-season adventure. The large YMCA outdoor pool was used, not only because it was convenient to the press, but also because hot showers were adjacent to the pool. This exercise almost ended in an unexpected way. Though ice diving sounds like a "polar bear club" event, it is not nearly as uncomfortable as one might think. If a reasonably mild day is selected, the surface activities aren't all that chilling, and the divers' exposure suits protect them quite adequately from the water temperature of 54-55°F usually found under the ice. As I mentioned in a previous chapter, the greatest attention must be given to provide a means for the divers to find the egress ice hole, since it disappears from view under water, unless the diver is directly under it. And, since it is virtually impossible to break up through ice that is much more than about an inch thick, the divers seldom wander more than a few dozen yards away from the hole. The large YMCA pool covered almost an acre, and was mostly quite shallow, except for a 12-foot deep 35-foot by 35-foot area under the diving boards. This is where we cut the hole in the 8-inch thick ice and commenced the dive, each pair of divers connected by rope to one another, and to a dive tender on the surface. The tender's assignment was to feed out and take in the line, keeping the slack out, and, if given a signal of three sharp tugs by the divers, he was to haul them in quickly.

In season, the edges of the deep diving area were marked with a rope, secured to the pool edge and encircling two steel posts on the outer corners of the area. While the rope was removed for the winter, the steel posts remained, extending about a foot above the ice. My buddy, Dick Spence, and I made our prescribed method of entry through the ice hole, and proceeded under the ice, up and out of the deep well into the shallower area of the pool. Visibility was much poorer than we expected, due to a layer of leaves that had been allowed to remain on the bottom of the pool at the season's end. These rotting leaves had colored the water so badly that it resembled strong tea.

About ten minutes into the dive, my regulator "froze up" and failed. (Today's modern regulators are far less prone to do this.) Luckily, it froze in the "open" position, so I did not suffer a shut-off of my air supply. Instead, the air suddenly began to flow violently from my mouthpiece. Not only did this make it difficult to keep the mouthpiece in my mouth, but the rate of flow would quite quickly deplete my air supply. Dick quickly grasped exactly what had happened, and we immediately instituted "Plan A," which was to give those three sharp tug signals to our tender. He, as instructed, quickly began to strenuously pull us toward the ice hole. Things were going well. With his pulling, and our own finning, we were moving at a rapid speed toward safety.

Then, because of the severely limited visibility, it happened that Dick passed on one side of one of those steel posts, and I unknowingly on the other. Being tied together with a 6-foot piece of rope, this of course brought us to an immediate halt, still about 15 feet from the escape hole. My air supply continued to blast from my regulator. Time was running out. I envisioned the headlines in tomorrow's newspaper: LOCAL SCUBA INSTRUCTOR DROWNS IN SWIMMING POOL. Our tender, thoroughly puzzled by the fact that he could pull in no more of our rope, simply kept pulling harder, enlisting another club member to help. This strain on our surface line made our predicament even tougher to resolve. Obviously, we untangled ourselves in time, and made it to the hole and the surface. (Otherwise, you wouldn't be reading this.) My air supply, by then, was totally expended. A discussion about this episode at the following club meeting concluded that it was simply a matter of "Murphy's Law" being in full force. The new club members who were on hand to make their own initial supervised ice dive chose not to enter the water that day. I gave them full credit for good judgment!

7

Search and Recovery Diving

Most every community with a river, lake, or ocean nearby has an organized group of divers to aid with underwater rescues or recoveries, in case of boating or swimming accidents, crimes, etc. In most cases, this group is known as the "Search and Recovery Team," or by a similar title. In many cases, this group consists of volunteers from the local diving club. In others, the team is made up of paid members of a local law enforcement agency or fire department.

In the case of our area, from 1961 to about 1970, members of our club volunteered for this work, and each one who did was appointed as a "Special Deputy" by the local sheriff's department. This provided the volunteers with a certain degree of liability protection, and even some "Workmen's Compensation" coverage, in case of an injury. (We were "paid" $1 per year.) This also lent some "prestige" to those who were selected to serve, since a jacket patch and card came with the appointment.

Deputized divers prepare for a search and photo shoot of a stolen car ditched in a local lake. (Circa 1975)

Now, in the majority of cases, a "search and recovery" team is most typically called on when an accident victim has drowned. Though the occasional ditched criminal's gun needs to be found, or a drowned car must be recovered, the diving team is most often used when a dead body must be found. Other methods of doing so, such as grappling hooks, etc., are looked on as far less thoughtful to the survivors, for one thing.

In our case, we were most often called to the accident scene at Buckeye Lake, a very shallow, mud bottom lake with zero visibility. Body searches there were exercises in blind groping, and not at all pleasant, to say the least. Nonetheless, new club members were often eager to be named to the dive team. Either the title impressed them, or they yearned for the status of Special Deputy. Naturally, it was the club's practice to accept new team members only after they had logged a good deal of varied dive experience. Even so, it often happened that new team members, when called to their first body recovery, developed sudden "ear aches" or other maladies that prevented them from diving that day. Even those who had participated in a number of previous practice recoveries under the same conditions would often develop such symptoms when a real victim was involved. I honestly couldn't blame them, and none of the other members ever ridiculed them, because all of us had experienced the same emotions. Particularly under these absolutely blackout conditions in this lake, feeling for a dead body was stomach-churning, to say the least.

The same fellow who I shared that pool ice dive incident with, Dick Spence, wasn't quite that sensitive to the new diver's qualms, however. Our underwater search patterns always involved a team of two divers. In that zero visibility, both divers were connected to one another with a buddy line. Dick absolutely delighted in getting paired with a new team member on his first body recovery dive. Midway through the dive, Dick would position himself in front of his buddy, and go absolutely slack. Remember that both divers can see absolutely nothing. He would, in effect, become a "dead body." When his buddy, with gloved hands, would then find one of Dick's "dead" body parts, Dick delighted in the reaction that ensued. Often, the buddy would feel up and down Dick's arm or leg just far enough to convince himself that he was touching a human body. More often than not, the newer diver, in terror of what he had actually found, would release his grip and head off in another direction, not able to accept the fact that he had actually touched the object of the search, a dead person. Dick, at this point, would usually lose his composure and have to surface, and the new diver was in for a good deal of ribbing for a while to come!

A classic situation once occurred at Buckeye Lake when a local middle-aged man sold his house. Having netted $40,000 from the sale, he ardently wanted to see the cash. Insisting that the bank round up $40,000 in currency, he stuffed the money in his pockets and called his best friend, who kept a boat at the lake. Intending to "do the rounds" of the lakeside taverns, the pair boarded the boat and headed out at full throttle for the other end of the lake, nearly five miles away, the newly "rich" man in the stern. The boat's owner, arriving at the first tavern destination and preparing to dock, was shocked to find that his friend was no longer in the boat! Clearly, he had fallen overboard, and he couldn't swim a stroke! But where? When our recovery team arrived on the scene, the sheriff could only say that the body was likely somewhere near the "middle of the lake, somewhere along a line nearly five miles long!" (The sheriff also told us about the cash, and that he hoped that it would not "have fallen out of the victim's pocket," once we brought the body to the surface!)

We searched for two solid days, groping all of the lake bottom we could cover. No body. Suspicions began to be voiced about the guy not being a "victim" at all, but maybe having skipped out on his wife with the cash. However, on the third day the man's body surfaced on its own. The cash was still in his pockets. And they say you can't take it with you....

On another body recovery, I had my own third experience with a "close call." Our team was called to the scene of a flooded river, two counties away. The spring flood was rushing over a concrete dam that was usually about six feet high, and the river's flow was such so that the dam was invisible, creating only a dip in the surface of the water as it passed over. A group of college students had decided to canoe the flooded river, and enjoy the "rush" of swooping over the fast ripple of the dam site. Three canoes made it easily, with no problem, but the fourth capsized, and the two young men were thrown into the river, not to be seen by their friends again. The local dive recovery team was called in, but the conditions were such that their leader determined that no diving operations could be safely conducted. This highly distressed the father of one of the victims, who was on the scene. He felt sure that the boys were trapped in the currents at the dam, and that they might somehow still be alive, perhaps in an air pocket under the downstream side of the dam. He insisted that the divers get into the water, *now*. They refused. This man was highly influential, and had connections to the Governor. Troops were called in, and a plan was developing to have heavy equipment re-route the river around the dam site, so the boys could be "rescued." A call also went out for other dive teams, and ours responded.

Once we arrived on the scene, it was immediately obvious to us that the victims would NOT be at the dam, and that the search should be directed downstream from there, where we were sure the boys (bodies) would be found. It was also perfectly clear that no dive teams should enter the water under these conditions, or more lives could well be lost. The father wasn't buying this, and he was railing about the "cowards" on the dive teams that were refusing to save the boys. Several fire departments were also present, with boats, but they, too, were reluctant to put boats and men into this raging torrent. Things were going nowhere.

I felt strongly that the operations should be moved downstream, and came up with an idea. I told the father that we had an idea for searching the foot of the dam. We would launch one of the firemen's boats well upstream of the dam, and lower it by means of a long rope to a point just slightly below the dam. We would then swing the boat over to our shore, two divers would board the boat, let the boat swing back out to the middle of the river, and, connected to the boat with safety lines, those divers could submerge and check out the dam. He went for the idea. One end of a safety line would be tied to the boat, and the divers would carry the coil of rope with them, to pay out as they needed it. The divers would be myself and another experienced recovery diver, that same Dick Spence. Dick and I made our *own* plan. Once the boat was in position, we would submerge, alright, but just barely under water, out of sight, just under the boat, where we would stay for about 30 minutes, reporting, when we surfaced, that we had made a search of the dam. Then, we hoped, once the father and the other authorities were convinced the boys weren't there, we could start searching the shore downstream, where we were convinced they would eventually be found.

All went well with the plan at the start. Enough rope was found to make the pivot point for the boat several hundred feet upstream, which worked quite well. The two firemen in the boat were clearly in no danger, while being lowered over the dam ripple. The boat was drawn to shore by another rope, we boarded, it was released from shore, and quickly swung on its pivot line to the center of the river. We submerged, Dick at the stern of the boat, and me from the bow. Dick's ruse worked perfectly. He simply "parked" under water, under the stern of the boat, as we had planned. The water was moving so fast that his exhaust bubbles didn't give him away. I, on the other hand, submerging about 20 feet closer to the dam than Dick had, was *immediately*, with great force, pulled directly against the downstream face of the dam, by the tremendous hydraulic reverse current of the water. This was bad news! Furthermore, in the suddenness of the force that gripped me, I lost the coil of rope. And, worse yet, the face of the dam was studded with sections of metal reinforcing bars that had been exposed by the eroding concrete. I was pinned against the dam, being buffeted around and against these steel bars, and my excess rope was quickly becoming entangled in them. This was a fix! I mentally calculated the chances of my extricating myself from this mess, and they didn't look good. "More headlines!" I thought. Determined to give it my best shot, I started to untangle and re-coil my rope, though I couldn't see a thing. How glad I was to finally fish it off of the last steel rod, and feel it tighten against what I fervently hoped was the end secured to the boat.

I pulled myself to the boat, needing all the strength I had, to fight the awesome pull of the reverse current. I made it, and was pulled into the boat, where Dick already sat, with a smug knowing grin on his face, thinking that we had pulled this off quite neatly. Of course, I couldn't tell him, at that time, what had taken place at *my* end of this operation!

Back at shore, we reported that we had searched the dam, and found nothing. Operations were immediately transferred to the downstream area, as we hoped. Secretly, I wondered if the boy's bodies might not, after all, be pinned against the face of the dam, like mine almost was. But if so, I knew they would not be alive. As soon as everyone's attention shifted from us, I began shivering like I have never done before or after. A combination of the experience and the cold gripped my whole body in shakes so severe that I couldn't hold a cup of hot coffee that was given us. All in all, it was not a wise decision we had made, and I was lucky to survive. Though conditions downstream were not much better for diving operations, divers waded the shoreline all through the night, aided by searchers in boats. Early the following morning, both boy's bodies were found, entangled in semi-submerged trees, nearly two miles from the dam.

When an emergency occurred, the procedure was that the sheriff's office would call a member of our team, who would alert the others. The sheriff was provided with a sequential list of who to call first, who next, if that person couldn't be reached, etc. The person who was reached would then be responsible to decide if the whole team should be called out, or only a few, or different "shifts" arranged, etc. Naturally, the guy at the head of this list was the guy who more or less considered himself as the team's "head honcho." Often, probably too often, he would call only one other diver, hoping the two of them could do the job. This, of course, would save others the trouble of leaving their jobs to respond (and, it would also serve to most likely get this guy's name in the newspaper report). OK, so some said he had a little "hero complex."

In the middle of a spring night, a flash flood descended on a local R.V. park, sweeping several of the camper vehicles downstream. Amazingly, especially because it happened so fast and in the middle of the night, no lives were lost. My phone rang about 2:00 AM. The sheriff was on the line. He said they had called out our team's "leader" about midnight, and he had shown up on the scene with one other diver. They had entered the water to see if anyone was trapped in a camper that had tumbled downstream, and now they, the divers, were stranded themselves, up a tree, mid-stream in the flood. I called five other divers at once, and we were quickly escorted to the scene, sirens wailing. Sure enough, here were our two "heroes" up a tree, about 30 yards from "shore." The current was still rampaging, and whole trees, cars, and more campers were being swept down the torrent. Our guys were clinging to the branches of a tree that, before the flood, were probably about 30 feet from the ground. I sensed that our "leader" would likely have rather spent the whole night in the tree, than to have to be rescued by his peers. However, their situation was indeed hazardous, and the water not yet crested. We commandeered a boat from a fire department squad, and, with ropes, and a diver in the water at the stern of the boat dodging floating debris, we worked it over to their "nest" in the tree. They plop-plopped into the bottom of the boat, and were pulled to safety. Grateful and relieved, certainly, but a bit sheepish at the same time. It was determined later that, in the future, no fewer than six team members would be called to respond to any situation, even if deemed "minor."

When called out to search for crime evidence, lost tackle, etc., and even submerged cars, we quickly learned that what we were looking for was very seldom anywhere near where the witnesses said it was. Of course, we always questioned as many eye witnesses as possible, but even when the location of the object was "pin-pointed," it would often be found great distances from there. Cars, particularly if their windows are up when they hit the water, can float surprisingly far. And, if a current is involved, even objects like guns, especially those with wooden stocks, can be moved along for hundreds of yards. And, of course, in the excitement of seeing a drowning or witnessing an accident, the excitement can often lead an observer to be far less than accurate in location descriptions.

Later in my career, I taught many classes in search and recovery procedures to law enforcement personnel. Never, I cautioned them, start a random underwater search based on someone's conviction that the object of the search is "right down there." The effective search and recovery team always starts with a planned, systematic search, often involving rope search patterns, buoy-marked sweeps, etc. The goal, after having X number of divers in the water for X number of hours, is to be able to know with confidence that the object is NOT in an area that has been efficiently searched, so that the search can be moved an adjacent area. Many specialized techniques exist to insure the coverage of every square inch of bottom, and this is how the search must be initiated. With the right equipment and procedures, a dime could be found in a ten acre field of straw. With random searching, days could be spent, and still no one able to say they know exactly where the dime is *not*!

Another "recovery" we made was not directly associated with our dive team, but was more of a salvage. Three of us were exploring a shallow wreck in Lake Erie when we found the large ship's propeller, still mounted on its 6-inch diameter shaft. For whatever reason, we thought this propeller would be a grand artifact, even though it was made of steel instead of brass or bronze. One of its four blades seemed to be buried in the sand bottom. First, we had to get the propeller off of its shaft. On our next visit to the lake, we towed a barge made of four steel drums, with which we hoped to float the propeller back to the mainland. To do this, we intended to flood the barrels, tie the "barge" snugly to the propeller, blow air into the barrels, float the whole thing to the surface, and tow it back. We estimated that the propeller might weigh as much as 1,600 pounds. On this trip, we also took a number of hack saws and several replacement blades. By the end of the day, our cut was hardly a third of the way through the shaft.

We sunk the barrels, tied them to the propeller, and left them, returning the next week end with more hack saw blades. We didn't cut clear through the shaft this time, either, and had to return again the following week. This time, we got it cut. As it fell to the bottom, we saw that the blade we had thought was buried in the bottom was instead broken off and missing. Even so, we soon found that our four barrels wouldn't float it. Back to the drawing board! The next week, we added two more barrels, got it to the surface, and laboriously towed it the five miles or so back to a marina dock, where we had it lifted out with a hoist.

Back home, we had it sandblasted, and it was pretty impressive, even with one blade missing. At the annual Diver's Council convention, our propeller easily won the "Best Artifact" contest, though we took some ribbing from some of the other clubs who chided us for not finding the missing blade. (We had searched the entire area and couldn't find it.) At the convention the following year, the diving club from Mansfield, our club's competitive "arch enemies," made a big ceremony of hauling in and presenting us with the missing blade, which they had managed to find and salvage. In this, they took great delight. In the meantime, our propeller had been mounted as the base of a flagpole monument in front of the YMCA, in a manner that now precluded the attachment of the fourth blade. The blade ended up being left at my house, where it gradually sank into the ground and was covered with grass. We've long since moved from that house. Someday, perhaps, someone might find that 400-pound propeller blade buried there. That should provide an interesting puzzle!

Postscript: In 1998 the Ohio Department. of Wildlife spent several thousands of dollars transporting the propeller **back** to lake Erie so "It could be seen (felt?) by future generations of divers."

A Chance to Help Advance the Sport

In some of the previous chapters I have described how the teaching of diving and the certification of instructors has been improved, through the years. A lot of very dedicated people devoted a lot of time and effort to this, and I was fortunate to have been able to work with some of them. This chapter is not intended to be a accurate "history" of these developments, but simply a relating of the progress as seen from my own perspective. Having spent but little time on the West Coast, where sport scuba diving first started to become popular, my viewpoints are mostly based on how things developed here in the midwest.

The Starting Point

Once "civilians" started scuba diving for recreation in California, the sport and the availability of the equipment gradually trickled east. Florida, of course, with its hundreds of miles of coastline, warm waters, and live coral reefs, was next to pick up on this new opportunity. Then, eventually, the word spread through the rest of the states that almost anyone could actually participate in this new adventure of exploring the unseen underwater world.

Early on, in California, it was seen that the development of some training standards and procedures were going to be required. After all, the scuba diver is operating in an alien world, sustained with a life support system, and subject to hazards and pressure-related maladies that present certain risks. Without some direction and regulation, the accident rate was almost sure to climb, as more enthusiasts joined the adventure. Not only would lives be lost and injuries mount, but the resulting bad publicity would likely curtail the sport, and/or bring in governmental regulation.

One of the first organizations that got involved in diver and instructor training was the Los Angeles County Parks and Recreation Board. At first, the program met a great deal of resistance to its rules and regulations.

Divers were already diving, and individuals were already teaching others how to dive, when this new entity came along with new requirements and enforcement standards. It could be said that divers, in general, are more-or-less individualists, and they didn't like to be told how to go about what they were already doing! Furthermore, there were going to be some fees levied, to help support the new program, and this, too, was greatly resented. However, with a lot of hard work and diplomacy, the real "pioneers" in this field eventually got a structured program working, and diving safety was immeasurably enhanced. In fact, they quite likely saved the sport from being regulated out of existence! Thanks in great part to the foresight of these early organizers, the sport has continued to maintain its own standards and outstanding safety record, requiring no intervention by law makers at any level, apart from those rules applying to fish and game regulations.

Here in the midwest, however, no such standards at first existed. Diving equipment began to be available by mail order from some of the west coast manufacturers and importers, and people inevitably began to experiment with it. While the suppliers made some printed guidelines available, and a few books on the subject began to appear, midwest divers were left to themselves to learn and experiment. Thankfully, at about this same time, the National YMCA Aquatic Program began to consider adding scuba diver training to their other outstanding programs. And, as I've mentioned earlier, when clubs and groups of clubs began to form, proper training tended to be identified as an early priority.

Diving Grows Up

It was about at this point that my buddies and I came into the picture. The scuba class that we took at the Toledo YMCA in 1959 comprised a total of about 15 hours, half in the classroom and half in the pool. No supervised outdoor open water dives were included. Today, typical basic scuba courses last for 25-35 hours, and additionally involve as many as five open-water dives under the control of the instructor and his trained and certified assistants. And, after this basic training, dozens of specialty and advanced courses are available to the new diver. This may at first sound like it is "harder" to become a sport diver today than it was in the early days. This is true to the extent that more hours of training are involved. However, today's training tends to be far less physically demanding than the earlier "military style" courses.

The added time spent in the classroom and the pool are spent not in more involved studies of diving physics and physiology, or more strenuous water exercises, but in learning the safe conduct of more comfortable diving. Somewhat more time is spent on learning more of the underwater environment as well, so the new divers are better prepared to understand the new world they are about to enter. And, in most courses, the students also learn how to be more conscious of protecting the underwater environment. Compared to earlier teaching methods, modern multi-media teaching aids also serve to provide today's scuba students with a great deal of useful input.

And, equipment refinements and developments have made the sport much more comfortable and safe. For example, I earlier described the exhilaration of the sense of "weightlessness" experienced by the underwater diver. In reality, in the early days, achieving a state of being of exactly "neutrally buoyant" was always rather difficult. A little too much lead on the weight belt, and one was bumping along the bottom. Too little weight, and one had to constantly fin downward to stay at a given depth. And, as the scuba air cylinder was emptied of its air, the diver gained as much as six more pounds of positive buoyancy as the dive progressed. The foam wet suits, as well, changed buoyancy with changes of depth, as their thickness was squeezed by the changing pressures. Now, with modern buoyancy control devices, which also double as emergency flotation devices, the diver is able to trim off his buoyancy to *exactly* neutral, at any depth, with the simple push of a button. This by itself adds tremendously to the comfort and ease of diving. New devices for "buddy breathing" also make it far easier to share one's air supply with a buddy, in case of an emergency, and automatic underwater computer systems now provide the diver with information regarding safe depth and time limits, especially when making repeated dives in the same day.

Without a doubt, the modern scuba instructor is far more professional and effective than was usually the case in the "early days." Certainly, the safe conduct of the sport depends on the training, and the quality of the training depends on the skills of the instructor. This is where all of the training organizations have focused, first L.A. County, then the YMCA, NAUI, PADI, and other certification groups.

Who Certified Who?

Early on, as I initially sought instructor rating, I benefitted from the ground work done by some of the pioneers in the early Ohio Council. Though my first instructor certification testing was brief and quite basic, by today's standards, it was "state-of-the-art" at the time. By the time I earned YMCA instructor certification, much more had been learned about training methods, but a great deal of latitude still existed in the matter of standardization. Meeting another instructor, we would usually ask, "Where and when did you get certified?" We would then base our judgement of the instructor's merit on the answer. For example, we might recognize an instructor certified at the 1970 Philadelphia YMCA Institute as having passed some killer standards, while someone who got through the 1967 Peoria YMCA Institute was known as being the recipient of a "gimme." At this time, all of this work was being done by volunteers. The faculty at any one certification institute could make their tests and requirements pretty much as easy or difficult as they wished. Some colleges and universities set up in-house scuba courses to offer their students, and these teachers were sometimes "gratuitously" certified as Scuba instructors by a compliant regional agency, with little or no control of their methods or standards. Some schools didn't even seek this "outside" certification for their faculty scuba instructors, feeling that the school name itself was adequate proof of training for the students.

All of this rather came into focus at the level of the dive shop, where the generally accepted practice was, and still is, to sell scuba equipment and refill cylinders only for those who can provide proof of training. The dive shop would have to develop its own standards as to which credentials were valid, and which were not. In any given day, the shop employee might be presented with dozens of different "C-cards," ranging from those of a local university; military UDT discharge papers; membership cards from a club from a neighboring town; YMCA cards; certification letters from a local "wildcat" instructor, etc, etc. And, the conscientious shop owners would also be aware that even "official" cards, such as the YMCA card, could have originated from a certain instructor known to provide inadequate training to his students. And still, at this time, many instructors were not requiring supervised open-water dives for their students.

In other words, while many dedicated people were trying to make this new sport both more readily available to the public and *safe*, the whole thing was going a lot of different directions at once. And, when it was seen that some profits might exist in the growing popularity of this new field, a number of charlatans inevitably appeared on the scene. Indeed, the sport was suffering some rather major "growing pains," and its future seemed uncertain at times.

Standardization In Instruction

As my own diving club joined with the state diving council and became more active in its projects, I became involved with the council's own training and certification program. And, once I had earned my YMCA instructor credentials and became associated with some of the people involved with that undertaking, I was invited to participate in the on-going development of the YMCA program. Early on, I became convinced that the greatest immediate need was not so much in the advance or changing of the standards that then existed, but in the standardization of the application of those methods. My goal, if I could find a way to be of any influence, was to achieve a point where a student taking a course in one city, from a certain instructor, would receive the same quality of training as would another instructor's student in a different city. Once this was achieved, I felt, we could then work on further improving the standards and the methods themselves.

In the process of working through the YMCA, I was appointed as the YMCA "Scuba Commissioner" for the Ohio/West Virginia region, which later expanded to include part of Michigan. In this post, my first act was to bring together a meeting of all of the individuals involved in the region's scuba programs, both the YMCA and the Ohio Council programs. This developed into a two-day "convention," held at the Mt. Vernon YMCA in 1966. It proved to be rather a landmark affair, involving participation by executives from the National YMCA headquarters, aquatic directors from several YMCA facilities within the region, and dozens of dedicated instructors. Principles from several of the diving equipment manufacturers were present, and helped to cover the costs of the event. By the time it was over, it became clear that we could all work together toward the same goals, the improvement and standardization of training.

Part of my (volunteer) job now became the approval of the criteria and methods to be used at instructor certification institutes. Finally, we could now develop standardized written exams, teaching skill evaluation methods, pre-requisites, pool skill requirements, etc. The staging of these certification institutes became far easier, and we were thereby able to make them more readily available throughout the region. As it developed, I was personally involved with the majority of these events, and made many new and valuable friends in the process. We also gradually lengthened the Institutes to two or more weekends, so that more training could be provided to the candidates. Clearly, it wasn't as "easy" (or cheap) to get an instructor's rating as it once had been, but I sincerely felt that the quality of the sport diver's training benefitted greatly.

Furthermore, we developed a means of enforcing these standards, so that an instructor who chose to teach in a way that fell short of the standards could be corrected and/or reprimanded. In fact, we insisted on the authority to withdraw the credentials from any instructor who repeatedly violated the standards. During my own term, this was only necessary a couple of times. I was, however, called on to "spot check" the quality of some instructors' procedures, from time to time.

In one case, I sat in an instructor's class while he described to his students that their bodies would get smaller and smaller as they descended deeper. He even sketched on the blackboard how their dimensions would be half of their original size at a depth of 33 feet, and further reduced by half again, at 99 feet I looked around and saw the incredulous look on the faces of the students. This fellow, somehow, had his physics kind of correct, but his effects all wrong. An air-filled balloon, taken down into the water, would, indeed, be "squeezed" to smaller dimensions, in the proportion he was describing. The human body, however, consisting of solid flesh and liquid, would not change dimensions at all, and even the air filled lungs would retain their original size, as the scuba regulator constantly and automatically pressurized them to exactly the surrounding pressure. At the end of the class, I asked to speak briefly to these students, and, as diplomatically as I could, corrected the misconception of the instructor. After the class, he and I had a little refresher course. He was quite well-intentioned and cooperative, but had somehow got his facts wrong on this particular matter. Still, he had logged some 100+ hours of diving himself, and I couldn't help but wonder how he could actually believe what he was telling his students!

Once we achieved the level of standardization we had hoped for, we still allowed the individual instructors to add to the basic teaching requirements, if they so chose. For example, at one point, YMCA instructors were required to provide at least 32 hours of indoor training. If a particular instructor wanted to make his or her own classes 40 or 50 hours, instead, that was acceptable. In fact, I myself always chose to exceed the minimum standards, in my own classes. A few instructors, however, persisted in teaching "Marine Corps" classes, forcing their students to pass strenuous tests that most of us felt were way beyond being realistic for the conduct of safe sport diving. When we discussed this with them, they made it clear that they felt the others of us were teaching a generation of "pansy" divers. More often than not, these instructor's own students would, if they were able to complete the course at all, fail to proceed with the sport. They simply had been turned off by the "Superman" image of diving that their instructor had put before them.

The Barefoot Instructor

Most of the better instructors would tend to present a professional image, dressing well for classes, being well prepared for the lessons, and always providing skilled demonstrations of the water skills to be taught. A few, however, chose to provide a far more relaxed and casual image. Often, at tropical island dive resorts, the resident dive guides would dress and behave in the most casual of manners, and this added, in a way, to the "vacation" aspect of the client's visit. No problem. Some instructors chose to project this same image in their classes back home, to make the whole thing seem more fun and informal. I personally knew a few who would teach their classroom sessions barefoot, dressed in well-worn "island" clothes. If the instructor was doing a good job otherwise, no real harm was done, but I do think that sometimes students who were professional educators themselves were sometimes taken aback by this informality. On the other hand, certain students seemed to relate to it.

Speaking of the "resort mentality," I am compelled to mention that, early on in the development of the sport, many resorts at dive destinations elected to offer "scuba training" to guests who spontaneously elected to try it. In effect, it often was a matter of "You got 50 bucks? I'll show you how!" Many of these abbreviated "lessons" resulted in the participants having very uncomfortable experiences, and being turned off to diving. The resort had the 50 bucks, but the guest who otherwise might have really enjoyed diving, if taught properly, was left with a bad feeling about it. Perhaps even worse, many people who did manage to survive this crude introduction assumed they were now "trained," and went on into diving at other destinations without really being adequately prepared. Often, such folks would come into our shops wanting to buy gear, sometimes thousands of dollars worth, and we were obligated to tell them that they must first sign in to and complete a certified course. Many got indignant at this point, but some would understand, sign up, complete a course, and afterward remark how glad they were that they had done so.

Today, most "courses" still offered at resorts are more properly identified as an "introduction" to scuba, and conducted in a more ethical manner, with the participants usually provided with a written recommendation that when they return home they enroll in a "real" course, to pursue the sport safely and with even more enjoyment. Still, however, there remain a few individuals at a few resorts that readily compromise common sense for the sake of some tourist dollars!

So Much To Learn

Eventually, I earned instructor rating with NAUI and PADI, as well as the YMCA and the Ohio Council, and sometimes participated with some of the other training agencies. There was a lot to learn about scuba, and the teaching of it, and I felt that it was good to "pick as many brains" as I possibly could. As I mentioned earlier, I was perhaps instrumental in the adoption of mandatory field trip dives by some of the agencies, as a requirement for certification.

Today, it is rather hard to believe that divers used to be certified without this valuable experience. There are those who say that today's new scuba student doesn't face the "tough" standards that used to exist. I personally feel that students now are far better educated and better prepared for safe diving, and every bit as capable of responding to emergency situations. (And, for that matter, far better at avoiding situations that could lead to an emergency!)

Today's instructor is taught that good teaching takes place when the instructor knows ten times as much about a given topic as he intends to get across to his students. For example, it would be theoretically possible to put together a two hour presentation on snorkels alone. Their history, evolution in design, the materials with which they are made, the manufacturing process, various ways in which they can be used and worn, etc., etc. But, with all of the other topics to cover, the instructor may have only ten minutes to spend on the concept and proper use of the snorkel.

The question the better instructor must answer is *"which 10% of what all I know about snorkels is going to be of the most use to this class, at this time?"*

Instructor Ethics

I do feel compelled to speak of a few "training agencies" that were formed for purely commercial purposes. As the sport grew, and sales of diving equipment expanded, there were those who jumped on the bandwagon without being loaded down with too many scruples. At least one of these organizations had the primary stated intention of getting new students "certified" as quickly, cheaply, and simply as possible, to therefore create a greater demand for equipment by as many new "divers" as possible. Potential students were "screened" for their realistic potential of purchasing gear. (Little profit is realized from the actual teaching of diving, and their philosophy was "why waste room in your classes on people who aren't likely to be buying gear?") Some classroom "training" sessions were dedicated to the measuring of students for custom-made wet suits, and the filling in of credit application forms. Dive shops belonging to these organizations were advised to target each student for a certain dollar level of equipment purchases, and then to turn their attention to newer students when this threshold had been met. In short, it was SELL, SELL, SELL. The quality of the training and the enjoyment of the sport was secondary, with these particular organizations. Some of this practice persisted well in to the '80s, but is now a rather rare exception, and, with the proliferation of more reputable agencies, rather easy for the discerning student to recognize.

Part of the psychology that allowed these predatory organizations to prosper is the almost infallible credibility that most students automatically attach to the instructor. After all, the majority of new scuba students know very little about the skills and mechanics of diving when they start their training. Whatever the instructor tells them, they feel, must be gospel. Whatever brand of gear HE uses must be the best brand to own. Here is a guy who has a lot of diving experience, who got himself certified to teach, and who clearly seems to know more about it than the students feel they will ever learn. Whatever HE says about diving must be true! Whatever HE says to do must be with good reason! Just like the traditional role of ski instructors, the scuba instructor can expect to find himself "idolized," to a degree. Certainly, he or she will be influential in the students' selection of equipment. That is all well and useful, IF the instructor is scrupulous and objective in his recommendations. If the instructor uses this influence to take advantage of the students, however, it is outrageous , in my opinion. That same concern has caused us to caution each of our instructors about allowing "social" relationships to develop with students of the opposite sex, during a course, or to even allow the appearance of a relationship. This is not because we are "anti-romance," but we have learned that fellow students can feel quite short-changed, if they find cause to think the instructor is somehow "favoring" a specific member of the class. Jealousy, it seems, is always ready to surface.

9

Making a Vocation from an Avocation

By 1967, I was teaching several classes each year at the Newark YMCA and at a few other YMCA and college pools in nearby towns. I was now being paid modest fees for some of these courses, but scarcely enough to cover the cost of the travel and the class equipment which I was gradually accumulating, piece by piece. Several members of our club were still helping with these classes, and often loaning the use of their own personal equipment as well.

Finally, the Newark YMCA Board decided to purchase several sets of diving equipment which we could use for these classes. The scuba program had now become a quite substantial part of their aquatic department. I was assigned the job of finding a source for this new gear. In the process of shopping around, some of my contacts in the instructional agencies led me to a source from which we could obtain the equipment at the wholesale, or dealer's, price. And, I learned that I could personally continue to purchase at this price level! Until then, when scuba students graduated from our courses, we had been directing them how to obtain mail-order catalogs for their equipment. Now, I could order it for them myself, provide them with a nice discount, and still be left with a few dollars of profit!

At about this same time, the Newark club purchased an air compressor, with which the member's cylinders could be refilled. Until then, all of the cylinders, including those we were using to teach with, had had to be hauled over to a dive shop in Columbus for refilling. The club raised the money to buy the compressor through a series of fund-raising projects, including car washes and Christmas tree sales. I was on the "purchasing committee" for this project, and again we found a wholesale source. Owning a compressor was a real boon for the club, and a tremendous help to our classes. There was, however, the problem of where to install and operate the compressor. The original intention was that it be "rotated" among the club members, being operated for about three months at each member's home. This proved troublesome, however, because the unit's motor required a more ample electrical supply than was readily available in most private homes.

In the end, with the agreement of my wife Alice, I volunteered to wire the unit in at the garage of our home, and keep it there on a semi(?)-permanent basis. The club would simply pay a per-fill rate, to cover the cost of the electricity. For each fill, I would collect a modest fee for the club from the members, so the club could cover the cost of the electricity, replacement filters, and maintenance. Since the use of the compressor system was available to members only, graduates of all of our classes were naturally quite eager to join the club, and it continued to grow. This all worked out quite neatly, and was certainly convenient to me, personally, as I could now very readily refill all of the cylinders I was teaching with! And, in the process of shopping for the compressor, and being in charge of its operation and service, I learned a lot about compressors, and this would prove to serve me quite well in the years to come.

I mentioned that I volunteered to house and function the club compressor only after Alice agreed. We both understood that she was likely to often be called on to operate the unit when members needed refills when I was out of town. This happened quite often, as I was still traveling for the meat packing company. Many of the club members themselves became quite familiar with the operation of the unit, and Alice usually only needed to open the garage to them. Even so, she still got a lot of practice at handling and filling cylinders! Only five feet tall, quite petite, and with three children to tend, this was a pretty big job, and, since each refill took about 15 minutes, her daily routine sometimes met a lot of interruptions, even though many of the club members were usually thoughtful enough to time their visits for times when I was most likely to be present.

Associated with the air compressor system was a bank of four large air storage cylinders which the club had leased from a local supplier of gases. This reserve supply of air could be used to somewhat speed up the refills, especially if a large number needed done at once. As a joke, I labeled these four cylinders OXYGEN;

NITROUS OXIDE; HELIUM; and NITROGEN. I would then ask "first-timers" how they wanted me to "mix" their refill, manipulating the cylinder valves as though I was actually flowing different of these gases into their cylinder. Most caught on to the joke early on. I took the labels off, however, when a newer diver was heard to complain of symptoms of nausea while breathing one of these exotic mixtures McBride had put into his tank! Talk about the power of suggestion!

One of the club members, Bill Coelho, was one of my most regular and skilled assistants. Other club members were also helping with many of the classes, but Bill was there for every one, and did an exceptional job, particularly with students who were having a difficult time mastering a certain skill. He clearly enjoyed it, and would later earn his own instructor rating. His cheerful personality and enthusiasm for diving was contagious to our students, and we became good friends. One evening in my office, as he was helping to prepare the paperwork for one of our class sessions, we chatted about the possibility of perhaps eventually making a part-time business of this whole enterprise. I was already counseling most of our students about equipment selection, and ordering for many of them. And, we certainly had a lot of contact with active divers, as they came to my home for air refills. Bill and I discussed how useful it would be to have a small supply of new equipment on hand, for "show and tell" and for better size selection, immediate delivery, etc. Eventually, we agreed to split the cost of a small inventory of gear, and experiment with the "business," as partners.

It was at this same time that a friend from Dayton, Bob Farmer, who I had met at my instructor Institute there, was killed in an auto accident. Bob had been operating a scuba regulator service and repair business from his home, and had all of the special tools, parts, and testing equipment required. As his was the only such service center in the area, he had established a substantial clientele. Soon after Bob's death, his widow contacted me with an offer to sell his regulator tools and parts. Bill and I felt this would be a fitting addition to our enterprise, and we dug up some more money to buy Bob's stuff.

Now, of course, we needed a place to function from. Since our house had a large full basement, it was decided to partition off a 12-foot by 15-foot section of it, and make a miniature "dive shop," (again, with Alice's forbearance.) Our regulator repair bench, which we built ourselves, took up almost a fourth of the space. We named our enterprise "Sub-Aquatic Associates." We worked out a schedule of duties that would evenly share the work. Bill and I worked on regulators in the evenings, and he did the new "company's" book-keeping at his home, to make up for the extra hours that I tended the "store." Having the club's compressor still located in our garage, of course, helped to bring a regular flow of divers into our little shop. In fact, my life was getting rather busy, now, as I was also still working a tough full-time job, teaching classes at least two nights each week, functioning student field trip dives at least 40 or 50 weekend days each year, and trying to spend some time with my young family. Our family "together time" was often picnics at the field trip dive sites.

Once we had a business name, and sort of a physical location, some of the major equipment manufacturer's salesmen that I had become acquainted with offered to sell to us at true distributor cost. This required an input of even more inventory capital, of course, but it meant that we could now also sell equipment at dealer prices to other instructors in other areas who, like we had started out, were mail-ordering equipment for their students. Now, all those friendships that I had made in my volunteer work with the various instructor organizations proved valuable, as many of these instructors began to order their gear from *us!*

Before we could feel confident with the servicing scuba regulators, Bill and I determined to attend the various manufacturers' service schools. Most of these were on the West Coast, however, and while it was a real stretch of our resources to arrange a trip there, we did so, spending a full day at each of several factories. We added another two days to the trip, and sampled west coast diving, which we found to be vastly different than the Caribbean! For our first dive we boarded a party boat at Seal Beach at midnight, for a seven hour trip to San Clemente Island, about 70 miles offshore. These boats, nicknamed "cattle boats," were designed to haul as many paying divers as possible. There were about 40 on this one. After about seven hours of rolling through a rough sea, nearly half of the divers were violently sea sick by the time we dropped anchor to dive. Even the sickest of them geared up anyhow, in some cases while they were laying on the deck and barfing. This astounded us, though luckily we got ourselves in the water before we succumbed to the epidemic. (Once in the water, especially near the bottom, which remains fixed, sea-sickness often disappears.) This is what these sickies were counting on. And, we learned later, everyone who boarded that boat (except us), knew full well that some 50% of the divers were doomed to get seasick. It always happened that way. They liked the diving so much, though, that they were willing to take the chance that maybe this trip, it wouldn't be them!

Back home, even with all of these projects going at once, we still found time to enjoy some of the lighter sides of the adventure. In our little shop one day, I was showing diver's watches to a recent course graduate, relating the various features of each of the models. "This one," I said, "is self-winding." The customer nodded his approval. "And this one over here," I remarked, "has a rotating bezel, so you can keep track of your dive time!" He was impressed. "And this one," showing him the most expensive watch, "even has a display for the day and calendar date!" At this, he looked incredulous, saying, straight-faced, "Oh gee, I don't really think I will be staying down for that long!"

We stocked a few models of spear guns in our little shop, for those divers who traveled to salt water. A customer was looking these over one day when another customer, Charlie White, was in the shop. I had been diving in the Florida Keys with Charlie a few weeks before, and on one dive we were seeking a nice grouper to have for that evening's meal. I had the only spear gun, a model which we sold in the store. Charlie spotted a nice grouper. Knowing he had never bagged a fish before, I handed him the gun, and, sure enough, the grouper joined us for dinner. Charlie couldn't have been more proud!

Now, when he noticed this customer deciding which gun to buy, Charlie pointed out the model he had speared his grouper with, and in all sincerity said to the customer "Every fish I've ever bagged I got with this type of gun, right here!" Of course, that's the one the customer chose, and I certainly wasn't inclined to challenge Charlie's emphatic endorsement!

Charlie, who was a short bundle of barely contained enthusiasm, often volunteered his help around our little store. Always ready and willing to do whatever needed done, he was sometimes hard to keep busy. My partner Bill, one evening, to provide Charlie with something to do, directed him to blow up an inflatable raft we intended to display. That proved to be an unfortunate choice of words, because Charlie attempted to inflate the raft with the air from a high pressure dive cylinder. Before he could get the valve closed, the raft indeed "blew up," rupturing like a giant balloon. Charlie was so distressed that he insisted on buying this raft, which we sold to him at our cost, and he patched it up and used it for a number of years. To this day, when we see Charlie putting air into his car or bike tires we call out "Remember Charlie, the word is INFLATE, INFLATE!"

Before long, the word spread of our little shop. Divers and "would-be divers" began to stop in. Some even traveled the 30+ miles from Columbus, where the shop there tended to charge rather high prices. Since it was our practice to require proof of certification before the sale of scuba equipment, we were obligated to ask each stranger about their training. The answers we sometimes received were inspired, to say the least. More than once we were told that the customer had a "long history of diving by breathing under water through a garden hose." (While this might at first seem plausible, it is absolutely physically impossible!) Often, we'd be told that the claimant had a vast background of diving while he was a member of the Navy U.D.T., "but lost his credentials." In truth, the Navy seldom issued certificates to its divers, so almost anyone could claim this service, and many did. In fact, one of my friends in the business once said "The Navy has trained 500 U.D.T. divers, and I have personally met 5,000 of them!"

Stanley Stemm, an experienced local club member, purchased a set of double tanks from us. Stan was a good-natured giant of a guy. He was in love with diving, but his air consumption rate was nearly double the average. His dives were invariably shortened, sometimes annoying his dive buddies who usually still had half of their air supply left when Stan's tank was depleted. Genial Stan determined to solve this dilemma by wearing TWO tanks. Now, two full-sized scuba cylinders weigh about 70 pounds out of the water, but Stan's size and strength made this of little concern. He had us rig these two tanks so that only one was turned on at a time. His plan was that once the first one got low, he'd know his dive was half over, and he would simply reach over his shoulder and open the second tank's valve.

Stan's "doubles" were the first in the club, and he was very proud of them. His very first dive with them was in a 50-foot deep limestone quarry in northwestern Ohio. His buddy was his newly-certified wife, Carole, who had a very low air consumption rate, compared to Stan. This was the first time either of them had been to this site. Their plan was simply to enter from shore, descend, and swim under water to a point on the opposite shore, about 300 yards away. There, they would surface, and decide where to scout next. As the dive started, Stan noticed at once that his tanks felt "lopsided" on his back.

As the dive progressed, this became more and more pronounced. Soon, one of the tanks seemed so heavy that it almost pinned him to the bottom! Halfway across the quarry, struggling and panting, he found himself literally crawling along the bottom. He ditched his weight belt, and even that didn't help much. Carole, swimming weightlessly above, couldn't imagine why Stan was behaving like a crab! Eventually, they reached the other side, where Stan clawed his way up the slope and dragged himself out of the water, gasping for air. Releasing the tanks, which now seemed to weigh a ton, he rolled over and laid there to catch his breath, muttering that these "doubles" clearly weren't such a good idea, after all.

Later, we discovered what had happened. Though he had had both tanks filled with air, the safety relief valve on one of the tanks had ruptured, unknown to Stan, perhaps while it was in his car trunk, releasing all of the air from that tank. As it happened, he had chosen to use the full one first. As soon as he submerged, the empty tank began to fill with water, through the now open safety valve port. Filling, it got heavier and heavier, soon adding about 40 pounds of negative buoyancy, plenty enough to put Frank out of balance and hold him to the bottom! Stan left his rig on the shore, walked back around the quarry, and drove his car over to pick up the tanks. Later, he brought his tanks into the store to have the one dried out and the safety valve replaced, and had us take them apart and make two single rigs out of them. He said that thereafter he would breathe all of the air he wanted, and when he ran out and his buddy had not, he'd simply go to the surface and get a fresh tank!

After a couple of years, our little shop proved far too small. At about this same time, my employer closed down their plant in Newark and moved the center of my operations to a new one they had purchased near the western edge of the state. This made my work days even longer. Bill and I decided to look for a commercial location for our growing business. Finding a little store in a modest rent area near Downtown Newark, we committed. I left my 20 year job to man the store. Bill kept his job, for now, worked evenings at the shop, and continued to do our book-keeping. I "traded down" our family car, and we scaled back our life style to necessities only, to handle what (we hoped) would be a *temporary* decrease in income. Alice had faith in our ability to make this work, and our kids, Shelley, Kelley, and Lori, were all excited about this new "adventure."

Our new store, though much roomier than the little shop in our basement, was actually just a "hole-in-the-wall" storefront on the ground floor of a four-story aged apartment building. It had previously been used for a succession of other businesses, most recently as a beauty shop. A small basement area was available for storage and for our service department. Actually, it was just

Our first dive shop, opened in 1963.

a cellar, and resembled a medieval dungeon, with a broken cobbled floor and walls of large, damp stone blocks. With each heavy rain, the entire area flooded with nearly a foot of dirty water that backed up through the storm drains. Everything down there had to be elevated from the floor, and we had to mop up after every rain.

The club allowed us to move the compressor from my garage to this new location, with the same arrangement as before, also granting us permission to sell air, on the club's behalf, to (certified) non-club members. This saved Bill and me a large investment, and the new location proved more convenient to the club members, as well.

Our little shop had one small rest room, which also doubled as our dressing room, for the fitting of swim wear, etc. The room had a unique antique exhaust system. An outdoor exhaust fan was connected by flexible duct work to a pick-up nozzle located just under the edge of the toilet seat. Clever and quite efficient, really! Early on, to be "cute," I placed a prominent label on the nozzle identifying it as our AIR COMPRESSOR INTAKE.

This got some chuckles, but we eventually removed it when someone took it seriously and reported it to the local Health Department!

Our sales and business grew nicely, both our retail and wholesale enterprises. And, on the side, we also started to sell high pressure compressors to local fire departments, who used the same type of compressor system to refill the air tanks worn by fire fighters. Eventually, this part of our business was to outgrow all the rest. And, we added a mail order operation for Speedo swim suits, selling mostly to swim teams. Our little shop was doing well. However, at about this time, Bill's employer offered him a promotion that involved a transfer to Georgia. He couldn't refuse, and we arranged a payment schedule that allowed me to purchase his half of the business.

Now, I needed some help with the shop, and hired a couple of part-time employees. Alice took over the book keeping. Several club members also volunteered some hours, in return for equipment discounts. One of the club member's employer transferred him to Columbus, and he offered to help us find a location there, and work part-time for us, if we would open a "branch store." We did, and it flew. About a year later, one of the major equipment manufacturers offered attractive credit terms if we would open a shop in or near Cleveland, where their brand was not represented. We did that, too. In another year, we were enticed to open another branch in Pittsburgh. Now, our little hole-in-the-wall shop in Newark was far too small, and we had several full-time employees!

I have a couple of memories from this era that are humorous only in retrospect. One Christmas evening, alone in our Newark store and getting ready to close up, I was surprised by a couple of customers who came in with a bottle of "cheer" to share. Together, we quaffed a few drinks and told a few stories. They left, and I gathered up our nightly cash deposit and the day's outgoing mail. I stopped first at a street-side mail box, then drove on to the bank. Reaching for the envelope to place in the bank's night deposit slot, I realized that I instead had the MAIL! I'd put the money in the mail box!

Nothing to do but to go to the main post office and beg someone to help me. Even on Christmas Eve, I found someone there, and even though I must have smelled of drink, the sweet man accompanied me back to the mail box, where I retrieved the money deposit. Maybe he was Santa Claus in disguise!

The other inglorious day was when I was running our Columbus store by myself. It happened to be my Birthday, and I may have been feeling a little sorry for myself for having to work alone for such a long day on this special day. (Poor me...) Late in the busy day, while refilling a customer's scuba tank which was immersed in a tub of water, the safety relief valve ruptured on the tank's valve. Not particularly a dangerous happenstance, the sound always makes one jump a foot off the floor. In this case, the blast of air blew all of the water out of the tub, and totally soaked a rack of several hundred boxes of swimwear. These had to be toweled off at once, or the boxes would be trashed. And, of course, the floor had to be mopped, and the whole area dried off. No one was there to help.

By the time I got this mess taken care of it was closing time. Finally! I hadn't had time to even eat lunch, and I was ready to GO HOME. In came a group of customers. Now, we needed every sale we could get, and I wasn't about to tell them to come back another time! An hour and a half later, they'd made their purchases and left. I quickly turned out the lights and headed for the door. As I slammed it behind me, I heard a great crash from within the store. Re-entering, I saw that an entire strip of fluorescent light fixtures, the full length of the store, had torn loose from the ceiling and smashed on the floor and onto some shelves of merchandise. Another employee was scheduled to open up the next day, and I couldn't let him be stuck with this mess! Two more hours spent cleaning it up. As I pulled from the parking lot late that night, I thought to myself, "Someday maybe I'll remember this day and laugh about it, but right now, Lord, just PLEASE let me get home!"

We decided to build a new building, with its own indoor pool, between Newark and Columbus. The pool design was quite unique, with a separate 16-foot deep circular pool joined to the regular full-sized pool with a four-foot diameter underwater "tunnel." By a fluke, we were able to secure a loan guarantee through the Small Business Administration. To come up with the down payment, we incorporated the business and sold shares in it. Retaining 61% ownership, we found among our diving friends fifteen gamblers who were willing to invest $600 to $6,000 each. Once this building was completed and opened, we closed down the stores in Newark and Columbus. As could have been expected, the building costs exceeded our estimates, and things were touch-and-go financially for the next couple of years. In fact, we couldn't afford to pave the new building's parking lot for about a year. Opening on January 2, the lot was a sea of mud, that first winter, and we found ourselves often pushing and pulling customer's cars out of the mire! Bless those customers who saw us through this period!

Our branch stores, in Cleveland and Pittsburgh, did quite well, at first. However, as expenses increased and more competitors joined the scene, they produced less and less profit. And, it proved more and more difficult to maintain capable and motivated management in both stores. One of our Pittsburgh store managers elected to room in the back of the shop, to save expenses. He also, against our objections, insisted on keeping a .357 Magnum handgun in the store, next to the cash box. He awoke one morning to find the store broken into and both the cash box and his handgun stolen! He had slept blissfully through the whole thing (and it's probably just as well that he did). By 1987, our son Kelley was serving as our general manager, and he engineered the closing down of our branch stores, selling one and simply closing down the other. By now, our compressor business, which we call BREATHING AIR SYSTEMS, had grown to encompass eleven states, and we had become the country's leading distributor for the most popular brand. Today, we also have compressor system sales and service centers in Illinois, Tennessee, and Florida.

In 1988 I received notice that I had been nominated for the "Small Business Person of the Year" in our local county. I was astonished! In the end, I not only won the county award, but also won out over 87 other candidates for the STATE award! This earned Alice and I a trip to Washington, where we met President Reagan, and contended for the national award (but lost out to Ben & Jerry, the ice cream kings). While I certainly appreciated the honor of the county and state awards, I was especially proud of the fact that the candidacy had involved a thorough investigation of our company's business ethics, which were found to be flawless.

The closing down of our branch stores left us with a bit of cash, and some under-utilized employees. We had also earned an excellent credit rating. Looking about for yet another enterprise, we ended up building a 500-seat banquet hall, just down the road from our other operation. This opened in 1992, and our experiences at making this work would likely fill another book.

The same year as my "Small Business Person" award, I was diagnosed with kidney and prostate cancer. Having both removed in 1989, I am now in good health, with a sound prognosis. However, this episode brought me to think about the energies required to continue to grow the company, so I began to concentrate on training others to do so. Our son, Kelley, is now our president, and doing a splendid job, and in 1995 I "retired" from regular office hours. Our little enterprise, started in the basement of our home, is now grossing many millions of dollars of sales each year. It has been an awesome experience!

10

Around the World with 800 Divers

One of the major side benefits of having been in this business has been the privilege of escorting groups of diving clients to exotic destinations all around the world. Dive shops, it turned out, were expected to provide dive travel opportunities to their customers, and we did so enthusiastically. Seldom a profitable enterprise in itself, "tripping" helps to maintain an interest among customers, and provides shop personnel an opportunity to travel at little or no cost, sometimes even being paid for the time. Most often, my wife and I would serve as trip escorts together. Thusly, we got to visit parts of the world that we could never have afforded to see on our own. In the process, we made numerous new friends of customers who would sign in to repeated trips with us. And, we made lasting friendships with many of the fascinating people who operated some of these remote diving facilities. As a result, we've visited almost every island in the Caribbean; several in the South Pacific and North Pacific; the Philippines; Mexico; Central America; the Red Sea; Australia; New Zealand; Cuba; and many, many other destinations. Along the way, we've had some real adventures. (Note that "adventure" is sometimes described as a tragedy you survive.)

The trip escort, of course, does have a number of responsibilities. It is certainly not exactly like a vacation. International flights and customs are usually involved, and it is the escort's duty to see that clients move through everything smoothly and comfortably, and that their luggage does likewise. Once at the resort, depending on the resort's own staff capabilities, the escort often is called on to serve as social director, side excursion coordinator, divemaster, bar companion, diet counselor, nurse, and, sometimes, therapist. In fact, our company actually has a manual, describing the trip escort's functions and procedures. The client's welfare, of course, always has to come first, even if the escort finds his own enjoyment compromised.

Typically, our travel groups number 12-16 clients. This, of course, means 12-16 different personalities, tastes, capabilities, tempers, sensitivities, and degrees of cooperation and promptness. Gradually, Alice and I found ourselves traveling with folks, mostly couples, who had been with us on several previous trips. Getting to know them, and their wishes, habits, and limitations certainly made our job a lot easier. As our business grew, other of our employees were given turns at escorting trips, and eventually Alice and I would lead only two or three major excursions each year, and the same clients often signed up for our trips. Often, it was almost as relaxing for us as actually being on vacation ourselves!

Diving In Communist Waters

One memorable trip was to Cuba, in 1979. For a brief period, Castro decided to experiment with the tourist trade, allowing in a few groups of divers, bass fishermen, and quail hunters. Apparently, the U.S. Government went along with this, at the time. Ours was one of a very few groups of divers who were provided with special visas for this adventure. Our group of 15 departed Miami on an American airliner granted special permission to fly to Havana. We were passed through customs at Havana with a good deal of ceremony (and confusion), then placed on a smaller Cuban propeller-driven "commuter" plane to the Isle of Pines, renamed by Castro as the Isle of Youth, just off the southern coast of the Cuban mainland. Here existed a sea-side resort, built prior to the revolution, which was being renovated for this tourist experiment. Just down the road from this hotel were the ruins of the vacation home of the former Dictator, Batista.

It was quite evident that our travels, activities, and accommodations were all being overseen by authorities high up in the Communist government, if not by Castro himself. However, all that was actually visible to us was the fact that our group had been assigned a married couple to serve as our "guides" and "hosts" throughout our stay. One might expect rather grim personalities to have been assigned this task. Instead, the couple, Julian and Maria Castansas were most pleasant, good-natured, informal, and even "ornery," to a degree. "Off the record," they confided, this was "the most enjoyable assignment they had ever been given." And, to our guests, they rather readily confided their disillusions with the Castro regime.

One day, after the day's diving, we were provided with bicycles-built-for-two, and allowed to ride into the countryside, which was mostly orchards and pasture rangeland. Julian and Maria, of course, accompanied us. Finding a giant Mango tree loaded with ripe fruit, we parked the bikes and sampled the delicious bonanza. Some in our group had never before tasted ripe mangos, and found them most luscious, eagerly eating a second, and even third fruit. Maria, we noticed, was really putting them away, commenting that this was her most favorite treat. Someone asked her "How many of these super-sweet mangos could one eat, before perhaps experiencing some bad effects?" Petite Maria responded that she "didn't really know, and had eaten 17 at one sitting once, without becoming ill!"

On our dive boat, five dive guides were provided, and they also proved to be jolly, fun-loving young men. Our meals at the resort were outstanding, and, we presumed, were specially prepared to create complimentary reports when we returned back to the States. About mid-stay, however, some of our guests began to yearn for some "good ol' American food," as often happens when visiting foreign countries. One of the guys in our group, Don Harper, operated a popular chain of pizza shops back home. One day on the boat, between dives, the conversation turned to pizza. It occurred to someone that it would be "neat" if we could all be allowed in the hotel kitchen that night, and, with Don's direction, all work together to whip up a batch of pizzas. As with all of our other requests, this was given serious consideration. The skipper of the boat got on the radio and notified the hotel chef of our wishes. A couple of hours later, the answer came back that it would be "O.K."

(Clearly, the request had traveled all the way to Havana headquarters and back.) The group was elated!

However, by the time we returned to the hotel later that afternoon, the permission had been denied. ("How would it look if guests had to cook for themselves," was the only explanation we could get.) "But," we were told, "not to worry, WE will fix you pizza for tonight!" And, indeed, when we were seated for dinner the staff began to bring us pizza, and more pizza, and more pizza. Amazingly, it was quite good, and we all stuffed ourselves. *Then,* they served us dinner, three full courses, followed by Baked Alaska for dessert! It was quite obvious that THIS was the dinner agreed on by the government, and was the dinner we were going to be served that evening, come hell or high water, and the pizza thing was just extemporaneous! We somehow felt obligated to do our best at consuming the dinner and the dessert, and could barely waddle back to our rooms.

The government, of course, runs everything in a Communist country, determining just how much of everything is going to be manufactured each year, etc., etc. As would any government, they habitually mess this up. One of the current problems, while we were there, was that the government had mis-calculated how much toilet tissue to make, and there was a critical shortage, expected to last for many more months. In public rest rooms, an attendant stationed at the door would dole out toilet tissue by the single sheet! And, it happened that the government had forgotten entirely to manufacture any toilet seats for the last few years, and those that broke could not be replaced. Of course, we all know that Fidel has a lot on his mind and can't be expected to think of everything!

The diving in Cuba was, by the way, absolutely outstanding. The window to tourism was only open for a few months, however. If the country is ever fully "opened up" again, it will be a major magnet for divers!

On our return trip through Havana, we were given a free day to tour the city. The most striking aspect was the vast number of American cars of mid-50s and older vintage. Of course, no newer American cars have been shipped there since the revolution. The ones that were there, on that date, have somehow been kept running, and, looking just about new! Other than these, all other cars are government-owned cars, mostly Russian and from other Communist countries. Otherwise, most of the population travels on foot, by bicycle, or on vintage motorcycles with sidecars, often loading entire families aboard. The city itself, other than a few government buildings, looks quite dilapidated. Mansions that must have once been grand are now divided into government apartments, with little pride or maintenance involved in their upkeep. Truly, communism has "leveled off" the social strata there, but has reduced everything to a sub-poverty denominator!

My Favorite Island

Bermuda is not exactly a tropical island, being much further north than most people think. Still, it has a semi-tropical look about it, and is absolutely the cleanest, prettiest island we have ever visited. I once asked a Bermudian how things there are kept so litter-free and freshly painted. He answered, simply, "We're PROUD of our island!" It helps, too, that a number of laws are on their books regarding the upkeep of automobiles and properties!

Bermuda is a shipwreck diver's paradise, with over 40 famous dive-able wrecks dating from as early as 1592. One of our favorites was the shallow wreck of the *Constellation,* a four-masted schooner wrecked in 1943. She was carrying a "general cargo," and divers still find all sorts of neat souvenirs among the wreckage. Nothing truly valuable, but trinkets ranging from cases of fishing lures to bottles of nail polish, medications, tools, and so forth. Indeed, it is rather like diving on the remains of an old-time general store. One of the most favored finds are sealed glass ampules of morphine. In fact, this was the "treasure" that inspired the movie *The Deep,* in which Nick Nolte and Jaqueline Bisset starred. Another ship, bound for the U.S. from Italy when she wrecked, still contains large stacks of Italian slate, destined to become pool tables. This slate, we were told, would still be perfectly usable, and is actually quite valuable, but not quite worth the cost of salvage.

On one trip to Bermuda, the only accommodations available were in what was then known as the Princess hotel, which was far more exclusive than the usual divers' digs. In fact, it was required that the 'gents wore coat and tie to dinner each evening. We complied, and generally behaved ourselves, but I must confess that we'd only brought along one tie and one dress shirt each, and by the end of the week the meals could probably have been counted on some of the ties.... After a long day on the dive boat and being under water, we were likely the most famished guests in the dining room, and plenty of "seconds" were usually ordered.

It was while we were waiting to board one of our return flights from Bermuda when we learned of the death of Elvis Presley. Many of those who were along on that trip remember the month and year of the "King's" death by remembering that 8/77 dive trip.

Meals To Remember

Another trip was to the Red Sea, with our destination being a hotel at Sharm-el-Sheik, near the tip of the Sinai Peninsula. It happened that a week before our arrival a massive midnight flash flood had roared out of the mountains and completely filled the first floor of the hotel with sand. In fact, a few guests were killed there. However, though the first floor was still buried, they had things working well enough for us to be accommodated on the upper floors. The kitchen and dining areas were not yet operable, though, so our meals were arranged through the nearby dive shop, and consisted mostly of frozen turkey, prepared dozens of different ways, mostly quite spicy.

Clearly, this was not going to be a gourmet's delight experience, but the diving was truly marvelous. By the second day, however, a serious situation developed. Unknown to anyone at first, it seems the flood had also contaminated the resort's water supply. Though we'd long since learned not to drink the local water at many of our destinations, we had been previously assured that the water here was "no problem," and we had all consumed some. The result was diarrhea of the worst kind, far surpassing anything like the famous "Montezuma's Revenge" of Mexico. SERIOUS scoots! Not significant cramps or nausea, just major RUNS, with no control. Of course, when one evacuates several dozen times each day, the exit becomes pretty darn tender. And, to top it off, we would face a dinner, each evening, of such as "Turkey Casserole a la' Jalapeno," "Peppercorn Turkey Soup" or such. Nobody missed any diving (this trip had cost a lot of bucks!),and we tried our best to make light of it, but we were all semi-crippled, to say the least. Some even found themselves needing to evacuate under water, during the course of a dive. The technique of doing this is not taught in the scuba class, of course, and it is not a particularly pretty sight, though done with practiced finesse it is not as disgusting as one might think. Luckily, the Red Sea water temperature did not require the wearing of wet suits.

Along this area of the Red Sea there exist a few popular nude beaches. Some of the folks in our group, between dives, found distraction from their bowel problems in scoping out these beaches with binoculars, from our dive boat. It proved rather amusing when our young males, eyes pressed to their binoculars, would first discover that the nude females they were ogling had a full compliment of leg and underarm hair. "Disgusting!" they would say, then stare some more. One could almost visualize the conflicts going on in their heads, as they viewed what, to them, was both enticement and revulsion in the same scene. I've often meant to ask a male from those countries in which female body hair is the norm just how "our" shaven females look to them. Perhaps just as unsettling as the reverse....

In Papeete, Tahiti, at the end of a diving day, on another trip, our hungry group of 18 descended on an oriental restaurant. We were all seated, amazingly, at one giant round table. The proprietor handed out menus, then stood back. Ten minutes later, no one had yet made a decision as to their order, while some were still asking details on this dish or that. Clearly, he saw, chaos was developing. He asked who the "leader" was, and I identified myself. He said "Lookee, you pay me $180 American, and I start bringing food of many kinds till you tell me stopee, O.K?" All agreed, and the feast began. Each dish brought out was exquisitely prepared and presented, and absolutely delicious! Everyone found several exotic dishes they loved, and the old gent kept on hauling out more till everyone pushed their chairs back. It was a dinner to remember! "Family style" meals are common, of course, especially in oriental restaurants, but this one was rather impromptu, and was fondly recalled by those who were present.

In another restaurant, another time, on another trip, in Auckland, New Zealand, our group was seated for dinner near the door to the men's room. It was a rather classy place. After we had ordered, I went to the 'gent's room to wash up. The urinal there was the fanciest I had ever seen! An ornamental stainless steel trough was provided with a shiny brass "shoe guard" on the floor. In position, the user's shoes would very effectively be protected from any splashes or splatters. Very civilized!

Returning to the table, I commented on this rather unusual arrangement. It happened that Pat Stover was with us on this trip, and she is a "trip" herself. Effervescent, always up to something, Pat is a real pleasure to travel with, and had a jovial "good ol' boy" type of personality. That is not to say she was at all masculine. Indeed, she is quite attractive. But ornery! When she heard about the urinal, she determined to photograph it. She asked one of the guys to go in to make sure no men were there, then entered with her camera, asking us to guard the door. Immediately, of course, a well-dressed gentleman from across the room headed for the men's room. Behaving just as we all knew Pat would have done, under the same circumstances, we didn't stop him. As he opened the door and stepped into the room, he faced Pat, whose camera strobe flashed at that very moment. He turned on his heel, and, with the most bland look on his face, returned to his table, never visiting the men's room the rest of the evening. He and his dinner companions watched as Pat came out of the rest room and returned to our table. They weren't laughing. In fact, they looked rather somber. We could only guess what he and they might have been thinking, and we couldn't bring ourselves to go over and explain the circumstances. Pat's spirits, on the other hand, weren't dampened a bit!

Treasure Pleasure

Another diving adventure was brought together in 1970, when we formed a group to explore a potential "treasure ship" off the western coast of Costa Rica. Dr. Irv Nickerson and his wife June were divers, and had built a vacation home on a remote region of the Costa Rica coast. They kept a boat there, and did some off shore diving, particularly around a group of rocky islands. Few others were diving the area at that time. Irv and June, one unusually calm day, dove near the shore of one of the islands, in an area that was usually too rough on most days. At a depth of about 60 feet, they spotted the remains of a shipwreck, obviously an old side-wheeler steamer. They collected several brass fittings, etc., and inquired, locally, what the name of that ship had been. It developed that the locals knew of no ship wreck to be in that area. Divers love to be the first to dive on a newly discovered wreck, of course, and they were excited. Upon returning to Newark with their souvenirs and underwater slides, we got together and investigated the possibilities. The design of the fittings and the fact that it was a side-wheeler seemed to put it in the era of the California gold rush, and the location was on the route of all the ship traffic from California to the east coast. Hey! Maybe our ship had come in — or gone down!

We mounted an expedition, timed with the best probability of weather patterns that might best provide favorable diving conditions over the wreck. Once there, one of our first needs was for provisions, and the trip to the small town of Liberia in Dr. Irv's Jeepster station wagon involved driving about fifteen miles on an uninhabited stretch of beach, since no roads led to their home. Approaching the mouth of a small stream, we found it to be at flood stage, and potentially too deep to drive across. Irv said that he'd faced such circumstances before, and that the answer was simply to drive around the mouth of the stream, a little ways out into the ocean, where a shallow sand bar would likely exist.

During an excursion to Costa Rica, our Jeep got caught by a rising tide while using a beach road.

Furthermore, it was now a period of low tide, and he saw no problem. He backed down the beach a ways, to get a running start, and headed into the surf. Things looked good for a while, but it was deeper than he thought, and the Jeepster stalled out. Now, we had a problem. Though the water was only reaching the bottom of the doors, the tide was now coming in, *and a tidal change of eight to ten feet exists in that area!* We clambered out with instructions to fan out and find a local farmer with a tractor or horses, to get that Jeepster out of the ocean as soon as possible. By the time we found one and got help on the scene, however, the water was up to the windows. We got it out before high tide, but it was drowned! We had the farmer tow it back to our base, and worked the balance of that day, and much of the rest, disassembling, drying out, and cleaning that entire Jeepster. We actually got it running again, though Irv soon after traded it off for another.

On this same trip, a member of our group was Juan Galvan, a Cuban expatriate. He volunteered to be our group's "interpreter," since few of the locals spoke English, and we had very little Spanish. As it turned out, his Cuban "Spanish" was nearly as different from Costa Rican "Spanish" as was our English. On the night of our arrival, we stopped at a local tavern for refreshments, and Juan volunteered to order for all of us, in the native tongue. I was the first to order, and told Juan that a Rum and Coke would be just fine. He turned to the waiter and, pointing to me, slowly said "r-r-rum and Coca-Cola," simply rolling the "*r*." I told him that this language looked to be a snap!

To make a long story short, we made several dives on the ship wreck, even using a powered air lift device I constructed, and found no "treasure" whatsoever, other than more brass and copper souvenirs and ornamental ceramic tiles. Later, when researching the wreck in the archives in San Jose, Costa Rica's capitol, Irv and June were able to learn the name of the ship, and the fact that it was simply a coastal trader, with a cargo of general merchandise. So much for treasure hunting!

Speaking of treasure, on another trip we had a group diving on the wreck of the *El Infante*, near the Florida Keys. This ship, *"The Baby,"* was part of the Spanish Plate Fleet, sunk in a hurricane in 1733. It had been salvaged, and much silver and gold removed. All that remained were timbers in the sand. However, sport divers were still sometimes finding the occasional Spanish coin or cannon ball among the timbers, and this was our hope. The method of searching was to simply fan the sand with a paddle, thusly excavating small holes in the bottom, where metal objects would sometimes be found. I picked a promising spot between two large timbers, where, I hoped, just maybe, nobody else had thought to dig. Sure enough, I soon began finding musket balls and the occasional piece of copper or brass. Very promising! I dug deeper and deeper, eventually reaching a depth of nearly my body length. Would my air last until one more swoop of my paddle uncovered that five million dollar gold necklace? Not exactly. Just as I was reaching the end of my air supply I saw the glint of glass. Wow — maybe a Spanish dish! I dug frantically, and uncovered a *modern coke bottle*! What a let-down!

Clearly, others had searched here before *me*.

Who's Afraid of a Big Bad Seawolf?

During a group trip to Rangiroa, in the Tuamato Islands, we participated in a shark feeding exercise conducted by Dick Johnson, who had been studying the sharks in that area for several years. Our group descended to a group of coral heads in an area about 40 feet deep, and each took up stations with our backs to the coral. Dick and his assistant proceeded to spear a number of snappers and trigger fish, and soon a number of grey reef sharks moved in, with an obvious interest in what was going on. We started photographing the action, which soon involved about 20 sharks, mostly around 5 to 7 feet long, and all getting rather excited. Next, Dick, who felt quite comfortable around this particular species of sharks, held the bait fish out to the sharks (using his left hand, since Dick is right-handed, and not **that** stupid!).

One by one, the sharks would approach, warily, and eventually overcome their caution and bite down on the fish offered. It became obvious that they wanted Dick to hold on to the fish, so that they could bite it in half. If one of the sharks found itself instead with a whole fish, they would bite it in two and let the other half drop to the bottom, where another shark would quickly grab it. This went on for 20 minutes or so, with the sharks becoming ever more agitated, but still behaving as Dick expected, though one grabbed his fish and darted between Dick's legs. Then, a couple of white-tipped sharks (not great whites), moved in to the area. Dick shut down operations at once, telling us later that the white-tips were far less predictable, and likely to also throw the reef sharks into a more erratic mode. We proceeded back up to the boat, each diver looking nervously behind him or her, and not dawdling on the boarding ladder. Some truly remarkable photos and videos were taken by the photographers in the group. We've since been on "shark dives" in other areas, but none have ever matched the excitement of this one, particularly since we were one of the first groups that Dick allowed to film this adventure!

Speaking of sharks, it has always been my habit, when diving with a group in salt water, to ask those who have just seen their first wild shark under water just what their reaction was. It seems that we all have a built-in fear of sharks, and in classes, new divers always have many questions about the danger of encountering a shark. This concern is best answered by the response I invariably get to my questioning, immediately after a dive when the newer diver has just seen a real, live shark. Each time, without fail, the diver describes an initial sense of awe and appreciation, at seeing this marvelous creature in its own element. "Were you at all frightened?" I ask. "Not at all!" is the certain answer, though a few go on to say that now, out of the water, they might experience a little "retroactive" nervousness.

It is, it turns out, a real treat to get to see these magnificent animals whose species has remained unchanged since long before man appeared on the earth. Clearly, they so inherently belong there, and the observer at once is brought to realize that he, the diver, is only a visitor. Is it likely that sharks will be encountered on most dives? Not at all. I myself had logged probably 200 hours in salt water before I saw the first one, and many of my dives had been in areas known to be commonly inhabited by sharks. On the other hand, some divers might see one on their very first dive! In any case, none of our trippers have ever been injured, or even threatened by a shark. Naturally, one would not want to aggravate them or entice them with such as a wounded speared fish, any more than one should bait or threaten a strange dog. Given normal circumstances, however, most sharks present little more hazard than an angel fish.

There are, of course, creatures in the ocean that can cause injury to the diver. In virtually every case, however, none of these creatures attack the diver (with the possible exception of the Great White shark, which is seldom seen in waters popular for diving). Sharp corals, stinging fish, needle-sharp sea urchins, and various other hazards dwell on the coral reef. In each case, the diver has to make the contact to suffer the injury. To repeat this, the diver has to go and *touch* the nasty critter, to get hurt. The answer, then, is simply to not go to and touch these hazards! It's that easy! If we see a diver exit a dive with punctures and welts, we know that diver did some boo-boos, and was clunking along bumping against things. The same guy would probably come home all scratched and dinged up from a walk in the woods! On the other hand, I've got a few punctures, stings, and burns myself, mostly from grabbing lobsters or getting too close to things while involved in close-up photography....

Well, There Was This *One* Shark...

Long Island, in the Bahamas, is nothing like Long Island, New York! We've taken many diving groups there, usually in the winter months, and sometimes other folks have asked "how in the world we can enjoy a vacation on Long Island when it's so cold!" confusing the two.

Several of the guys in one of our groups diving there were game oriented, and the owner of the resort offered to boat us to Conception Island to replenish the hotel's freezer with grouper and snappers. This was about a 70 mile round trip, and the small island is uninhabited. Its waters were seldom being dived, at that time, and fish were plentiful. By Bahamas law, however, spearfishing could be done only while breath-hold diving, without scuba. And, spear *guns* are not allowed, only hand spears, or Hawaiian slings. Most of those in the group who were interested in hunting were reasonably accomplished at this. One particular area was known for large groupers, though it was about 50-60 feet deep, which was about our limit. By the time one submerged that deep, found and speared a fish, and traveled back to the surface, a breath of air was starting to seem really yearned for, to say the least! This was particularly true because groupers are seldom speared while free-swimming, but instead tend to "hole up" in crevices, and often need to be wrestled out, once speared. In fact, this is why the better hunters learn to try for a kill shot, and the grouper's head is marked in a way to provide a distinct target zone for a spear shot that kills instantly, if one is that accurate. However, this is a head-on shot, and a miss usually misses the fish altogether.

I had harvested a couple of nice fish, and sent them back to the boat with our "runner" on the surface. The job of collecting fish from the hunters and swimming on the surface back to the boat with them was not a choice occupation, particularly in "sharky" waters, as these were. We usually assigned this task to a newer diver, assuring him that if he hurried back to the boat with these fish there would be no problem. They seldom had to be told this twice!

As I had speared the second of the fish, I noticed a grey reef shark, slightly bigger than myself, observing my actions from several yards away. Now, as I descended back to the bottom and was looking for another, this guy joined in the hunt. I'd look in a hole in the reef, then *he'd* look in the hole, apparently to see if I'd missed seeing something. Finally, I found a medium-sized grouper, and made what turned out to be a poor shot. He struggled deeper into the hole, making a lot of noise, and I couldn't get him out. My spear eventually pulled out, and I was by now more than ready to head for the surface. Just before I pushed off from the bottom, I noticed a nice helmet shell, and picked it up. I also noticed that Mr. Shark quickly looked into the hole where I had obviously wounded a fish. As I finned toward the surface, I looked down at the shark, who was now looking up at me. He saw that I had something in my hand, perhaps thinking it was a fish. He headed directly for me. I still had over 20 feet to reach the surface. I released the helmet shell, and the shark immediately snapped it up in his (her?) jaws, stopping to chew on it. In short order, the shell was spit out, and the shark gave me a dirty look, as if to say "What a lousy hunter!" and headed back for the bottom. I chose to move to another area to hunt.

Down Under Down Under

On a group trip to the Great Barrier Reef in Australia, we experienced a "double-cross" that ended up leading us into some unusual diving opportunities. Through a west coast travel agency, we had booked our flights and a fine live-aboard charter boat. We landed in Melbourne, then flew on to Cairns, arriving in the evening. Staying in a hotel that night, we were to board the yacht in the morning. I had been told to call the skipper from the hotel that evening, to advise him of how to stock the ship's bar, in case our clients had any special preferences. When I made the call, the skipper hedged. He said "Look, why don't you come down to the dock, and we'll talk about it there." Uh oh! I insisted that we were tired from the trip, and all that was necessary was to see that we had some good bourbon and rum on board. He repeated his insistence that I "come down and see the boat." I went. They had chartered our fine cruiser out from under us to a rich group of Texans that wanted it for twice as long and who would pay more. They had (how nice!), arranged this "swell" other boat for us. It happened that our substitute boat, however, had no guest cabins. Instead, they had bought a pile of bed rolls and camp gear, and intended to put our group ashore on a different island each night to sleep! Now, that would have been a fine adventure, if one enjoyed camping, as I would, but it was not what these clients had paid for or expected. Most, though nice enough people, weren't exactly the "rough it" type! I told the skipper to off-load the sleeping bags.

We had no choice, at this point. I called back to the States to the travel agent, and got them to approve a hotel in Cairns for our nights, and a promised partial refund for the clients. Next, we had to find a hotel (the one we stayed in our first night there had no un-booked rooms). Luckily, we found one, with a most congenial owner, and a bus to transport our group to and from the docks each morning and evening. The problem remained, however, that the barrier reef is located 25-40 miles out from shore. This would be a two hour trip, each way, morning and night, to and from the dive sites. It would also limit the areas we could travel to, compared to cruising on a live-aboard and anchoring each night at various sites out on the reef.

As it turned out, our "new" boat, the *Melawondi,* was, indeed, a fine day boat, and the crew was most congenial and helpful. It was roomy enough that the divers could, if they wished, catch a good nap on the way to or from the reefs. Still, it made for some long days, and left no time to explore the town or anything else on shore. Returning to dock late each afternoon, it was find some food, hit the sack, and be ready to crawl out at daybreak the next morning. Not exactly the luxury trip the clients had expected. Still, this group proved most good-natured and understanding, and not a one of them got sour.

The side benefit of this arrangement was that our skipper knew of some dive sites that were not regularly visited by most tourist divers. We visited areas populated by giant Tridacna clams, many weighing 600 pounds or more. These were the famous "man-eating" clams one reads about, though none of our divers ended up as clam food (man chowder?). And, Skipper Randy located areas lush with abundant spectacularly-colored various types of soft corals, and our photographers had a field day. All-in-all, the clients agreed that we had been privileged to see some of the very BEST areas of this famous reef. The Great Barrier reef has always been a magnet for divers from around the world, and many yearn to be able to say they've been diving there. In reality, in my opinion, it is unique mostly in its *size*, stretching for more than 200 miles. Many segments of it are, frankly, not all that different from coral reefs that are not halfway around the world. On this trip, however, thanks to the expertise of Captain Randy, we got to dive in areas that were truly extraordinary! And, at no time were we ever even within sight of other dive boats, which lent even more to our sense of exploration! Bloody fair dinkum, mate!

Come Fly With Me

Group trips to the Bahamas "Outer Islands" were often an experience in terms of getting there. Though jet-liners fly to Freeport and Nassau, the other islands are served by smaller "commuter" planes. Island air carriers tend to change around quite a bit, and service is sometimes erratic. And, the air strips on some of the smaller islands are often primitive, to say the least. One of our more memorable adventures occurred on a minor island "airline" I'll call BAHAMAS FBN. (FBN being Fly-By-Night). From Ft. Lauderdale our group of twelve was to fly to the Stella Maris Inn, on Long Island, a trip of about 375 miles, or three hours.

We arrived at the FBN ticket counter just in time for the 1:30 flight. We were advised that the plane had not yet arrived, and that a delay would be involved. Finally, at about 4:30 we and our luggage were loaded into a 12-seat twin-engine, high wing, propeller-driven Fokker. The pilots cabin was not separated from the passenger compartment, and we could hear their conversations and converse with them. We knew that the runway on Long Island was paved, but not lighted, and had no radar. At this time of year, darkness fell at about 7:00 PM. This was cutting it close! Then, instead of taking off, we taxied to a fuel pump, where another 45 minutes was spent. When we finally took off over the coast, the sun was getting mighty low. By the time we passed Nassau, it was dark. The Nassau control tower radioed to our pilots that a DEA Officer was in the tower, and wanted to know who the hell we were, and where we thought we were headed. We were instructed to land at Nassau for an inspection and an overnight. Our pilots frantically radioed their office and the Inn at Long Island, and eventually got permission to proceed. The plan was that the staff at the Inn would *light the runway with car lights* for our landing! Circling around Nassau in the dark while this was all going on wasted another 30 minutes.

Now, we were headed out over the black ocean to a destination in the dark. The pilot and co-pilot were frantically poring over charts with a flashlight, and, to our dismay, we learned that neither of them had ever flown to Long Island, even in the daylight!

Two in our party were non-diving middle-aged wives. We were all getting nervous, as we flew through the darkness. These two ladies, particularly, were getting more and more tense. Then someone remembered they had brought along a bottle of rum, and it was fished from the baggage stowed behind us. Appropriate doses were passed around the cabin, along with half of a sub sandwich passed back from the pilot. Things mellowed out considerably. I felt obligated to assure everyone that things were absolutely "cool," but I couldn't avoid feeling apprehensive. After all, if we missed Long Island, the next stop would be Africa! I stared down to the black ocean, hoping to see lights I could recognize as known islands. I had flown this area dozens of times before, and in the daytime, I could find my way. Nothing. Now and then a single light could be seen, but these could have been ships.

Finally, to our great relief, the engines were slowed and we began a descent. Ahead, through the windshield, we saw lights. Blessed lights! As we grew nearer, we could make out the runway, dimly illuminated with about a dozen car's headlights. As we touched down, everyone broke out in a cheer, including the pilots. We were rushed past the customs agent, and taken to the Inn, where a late dinner had been prepared for us. In the following days, as we traveled around the island, we were several times asked by islanders if we were those folks who had landed after dark the other night. It turned out that it was such a rare occasion that several folks had turned out to watch. We were certainly glad they saw a safe landing!

Another flight, another time, on an El Al jumbo jet from Israel, we experienced another incident, this time an adventure in rudeness. Early in the evening of the flight we were served dinner, by the not-very-friendly staff. Immediately after dinner, the entire serving crew retired to a section of seats where they played cards, conversed, and totally ignored the passengers for the entire remainder of the several hour trip. Once pulled up to the arrival gate at Kennedy, and as the passengers started to collect their belongings from the overhead compartments, the staff refused to open the doors until absolutely every passenger was seated. "Sit DOWN, everyone, or the doors won't be opened!" Ruffled, the passengers reluctantly complied. The staff collected by the door, opened it, and scrammed, every one! The passengers were left to themselves. That wasn't NICE, and there, now I've told on them!

The World Series, Mexican Style

During a series of dive trips to Cozumel, Mexico, several of our regular clients became well acquainted with the serving staff at the resort's dining room. Most of the staff were of Mayan descent from the town of Merida, in the Yucatan Peninsula. All were jolly and fun-loving. In our rooms at the resort we had no radios or television. We were at supper one evening when the Cincinnati Reds were scheduled to be playing in a crucial game in the World Series. Several guys in our group were real Red's fans. The waiters, learning of this, told us they had a radio in the kitchen, and could keep us posted on the running score. Every now and then one would come bursting from the kitchen and tell us something like "The Reds just got a home run, and are leading the game!" Then, later, another would come out with a long face, and tell us the other team had pulled ahead. By the time dinner was over, the Reds had won, our guys were exuberant, and the waiters, who all professed to also be Red's fans, joined in the celebration. It was the next day, when the newspapers from the States arrived, that we learned that the whole game had been rained out, and the waiters had been making the whole thing up for us! What an act!

Avoiding the Customs Tax

At least once each year, through the '80s, I escorted groups of college age kids on diving trips to the Florida Keys and to various destinations in the Bahamas. These groups were usually comprised of 14-18 youngsters of both sexes. Often boisterous and exuberant, these kids were nonetheless a lot of fun, quite meticulous in their adherence to safe diving rules, and reasonably obedient to my constraints.

Returning from a week of live-aboard diving around Grand Bahamas Island, one of our college groups was awaiting our flight departure from Nassau. Several of the guys had purchased liquor and rum from the duty-free shop there, without checking on the limitations involved. Once we entered the U.S. Customs clearance area, three of these characters learned that they each had bought one more bottle of booze than was allowed to be duty exempt. They could bring it through, but they would have to pay a heavy tax on it. This wouldn't work, for two reasons. First, they had spent the last of their money on the booze, and, second, they weren't about to pay any dumb old tax even if they had the money! Some of our boat crew was at the airport to see us off, and I suggested that the spare bottles could simply be given to them, as an extra "gratuity." No way! These three had bought the booze, and, by God, they were going to drink it! We had about a half hour till our plane boarded. Despite my pleas, the three yahoos went out onto the airport lawn and each chugged a whole bottle of rum, the whole while singing rowdy pirate songs.

By the time they got back to the customs area, they were bordering on belligerent, taunting the customs agents about trying to tax what was now in their bellies. Needless to say, once the plane was in the air, all three of these guys deposited their "tax-free" rum in the commodes! By the time we landed in the States, theirs were the most sober faces in the group. Even so, they were thoroughly convinced they had done a most clever thing indeed!

The Dolphin Act

It is a very rare treat indeed to swim with dolphins in the wild. We often count ourselves lucky to even get to see them from a boat.

On the very last dive of the very last day of a group trip to Cozumel, Mexico, when all of our divers were in the water at the same time, a whole group of dolphins joined us, and swam joyfully with us for at least a half hour. What a marvelous experience! These were the first we had seen all week, and they suddenly just appeared from nowhere. There were several whole families, including several youngsters. They "chirped" and squealed to us, cavorted with us, allowed us to touch and rub them, and just generally made us welcome to their world! Back aboard the boat, our guests were ecstatic, and joked with me about my "cleverness" at arranging this highlight for the very last dive. The boat crew, likewise, could not believe our luck.

We revisited Cozumel about eight months later. A few of the same clients were in this group, and they teased me about how I was going to again arrange the "dolphin finale" for the week. Sure enough, once again, on the very last dive of the last day, we were joined by a large group of dolphins! The divers couldn't believe it! (Nor could I!) Again, the dolphins mingled with us and hung around for almost the entire duration of the dive. Fantastic!

After this dive, some of the clients who had been with the dolphins on the previous trip seriously began to question me about how I had worked this out. There was no way I could have, of course, as it was sheer luck. Still, upon returning home, some of the clients spread the word about "Bill's trained Cozumel dolphins," and several of them signed on to our next trip there.

At our pre-trip briefing, I told these guests, and several others who had joined the group in anticipation of getting to swim with the dolphins, that what had happened on the previous two trips was purely happenstance, and that they should no way expect it to be repeated. After all, we'd been going there for years without seeing dolphins under water, and, so far as I and the boat crews knew about, dolphins just never behaved like this in those waters.

To make a long story shorter, for the last dive of the last day of this trip we anchored near where we had seen the dolphins previously, entered the water, and within ten minutes were joined by a whole circus of playful dolphins! Unbelievable! The boat crew swore they had not heard of even a single dolphin approaching a diver since we had been there last, though they often anchored hopefully in this same area. Even THEY wondered what "magic" I had worked, to call in the dolphins. My reputation was made! Smart enough to quit while I was ahead, I assigned other employees to escort our Cozumel trips thereafter. Never was another dolphin seen under water during any of these trips. Would we have enjoyed another magic experience if I had gone back? We'll never know....

Confessions of a Collector

Today, we all know that the underwater environment must be protected, or it will decline in the same manner as has so much of our top-side habitat. In fact, most of today's scuba courses include training on diving in an "environmentally conscious" manner. In the early days of diving, however, few gave much thought to this. The oceans were so large, it seemed, and the divers so few, that surely we could do no lasting harm. We took specimens, dropped anchors on coral reefs, and stripped wrecks of anything that could pass for a souvenir. When finally we began to wonder if all of this was really wise, our next thought would usually be something like "well, it really would be kinda' neat to leave this porthole on the wreck, for future divers to see, but if *I* don't get it *now*, the next guy that dives this wreck *will*." And, once we started to ponder the wisdom of removing live corals and shells for home collections, a visit to a tropical shell shop would make our "conservation thoughts" seem futile. Gradually, we've wised up.

Diving trips to the Yucatan Peninsula, in Mexico, used to feature dives for black coral. Dive guides were eager to guide customers to the best collection areas, in hopes of receiving nice tips. These same guides would readily harvest sea turtles, completely wipe out all of the lobsters and game from whole areas, and generally behave in the most classic short-sighted manner. Black coral became harder to find, and divers had to go deeper and deeper to collect it. Eventually, it became clear to all that, before long, black coral would no longer be available to attract divers to the area. Even so, if the guides knew of a remaining patch, they would take divers to it. The theory again, I guess, was that if they didn't collect it, someone else would!

It was at about this point in time when one of the guides asked one of our groups if we wanted to get some "really nice" black coral. Now, I had never collected any black coral, not because of any particular principles, but it just didn't seem to interest me that much. But, shame on me, I had, for a long while, thought it would be nice to have a stem of black coral large enough to be able to make finger rings by cutting cross-sections of it. When this guide, Pedro, seemed eager to take us to the coral, I asked if it might be large enough for this. "Si, Señor Bill, this is VERY large black coral!"

"But," he said, "it is very, very *deep*, and I wouldn't take just *average* divers to it.." That did it! I was to buddy with Pedro, and another couple, Bill and Nancy Ginn, elected to accompany us, and observe the operation from a shallower depth.

I will not mention the depth of this dive, because it was well beyond the "stupid" range. We had a maximum of 5 minutes to get there, hack-saw through the trunk of the black coral tree, and start back to the surface. We knew we would be affected by nitrogen narcosis well before we even got to the coral, and agreed that we would "keep an eye on each other." (That's rather like two drunks looking out for one another.) We began our rapid descent. Pedro had the hack saw, and headed directly for the tree. By the time I got to the bottom, I was totally euphoric. My bubbles seemed to be making the loveliest tingling music, I was completely "at home," and I felt no urgency whatsoever in returning to the surface. Faintly, I could hear the symphony of Pedro's sawing, and I hoped that this could last forever. In fact, I had left Pedro's side and started to descend even deeper down the slope, where things might even be more "heavenly." Then, my big toe saved me! All week, I had been suffering with a throbbing ingrown toenail. I had even had to cut out a hole in my fin pocket, to relieve the pressure. Now, just as I was inclined to sail away into the "deep blue," this old toe started throbbing again. Hard! Ouch! Damn! It brought me back to reality. I swam to Pedro and we wrenched free the half-sawn tree and headed to the surface with it. It was a dandy. I still have parts of it. But, aboard that boat was a group of people who I had trained, and who I had warned repeatedly against the stupidity of ever going as deep as I had just been! I was castigated, admonished, chided, rebuked, reprimanded, reproached, and put down in all kinds of ways that aren't even in my thesaurus. Pedro simply went below deck, to get out of the storm of criticism. I wanted to join him. I've never believed in the concept of "Do as I *say*, not as I *do*." Making this dive was an error in judgment, and was an extremely poor example. I lucked out. My toe saved me.

The Greenies

Diving involves a good deal of boat travel, and boat travel sometimes involves somebody getting seasick. I've only suffered the malady a few times, but enough to know that it is absolutely no fun, and that when it's happening one tends to think it is probably something else, probably something far more serious. When fellow passengers get seasick, I have, through the years, learned of a few things to help others avoid the onset, or at least lessen the suffering. I have also learned that humor is absolutely no help. I almost learned this the hard way. Returning with a dive group from a dive in Georgian Bay, Canada, the sea became quite rough. The boat was a semi-converted fishing boat, totally enclosed, and still smelling of fish. Our group of about 20 was standing elbow to elbow in the dim and thrashing cabin, and, one by one, turning green. To a young lady who looked about to up-chuck, I made a light-hearted remark, intended to perhaps get her mind off her predicament. She glared at me, turned to her husband, and instructed him very explicitly to "go over and hit that son-of-a-bitch in the nose, *hard!*" He didn't, thankfully, but I have never since tried to joke anyone out of this plight.

On a rolling boat, getting ready to dive, if someone is starting to feel the early symptom of seasickness, the very worst thing to do is to put one's tank on, then stand on the deck for any length of time. The weight of the tank on the back is often just enough off-balance effect to really bring on the barfs. The best answer is to get off the boat and into the water, fast! Preparing for a dive in the Cayman Islands one day, we were in a choppy sea. My partner for this dive was an attractive young lady, still rather new to SCUBA diving. Once she was nearly ready, I entered the water, first telling her to follow me as soon as possible, as I could see light shades of green flushing across her face. Once in the water, I waited for her, but she didn't enter. In a couple of minutes I surfaced, just beside the boat, to learn the reason for the delay. As I surfaced with my mask facing up,

I came face-to-face with her, leaning over the rail. In an instant, I knew exactly what was going to happen next, and in another instant it did. She chucked her breakfast right in my face. Hurled. Barfed. And, she had eaten a nice large, colorful breakfast. With my mask on, and my mouthpiece in, it was not that big a deal. I ducked under water, and it was gone. But it was a moment to remember, and her apologies, later, were most touching.

My Dream Island

Truly, this traveling around the world has provided us with invaluable memories and unique experiences. On one trip, we anchored off shore of an almost perfectly round small island in the Mindanao Sea, near Cebu, in the Philippines. In the center of the island was an ancient working lighthouse. Though we were told the island, called Balicasag, was inhabited, it was totally surrounded by reefs. No port or docks existed. Supplies were hauled in to the beach only by dug-out canoes or the unique one-cylinder engine powered *bangkas* ("bonkas"), which were basically over-sized canoes equipped with outriggers. My dive buddy and I determined to snorkel into the island. Our skipper suggested taking along some cigarettes, in a plastic bag, to give to the islanders.

Approaching shore, we saw that a large group of excited children had gathered on the beach, with a few adults sort of standing off behind them. We walked out of the water, with the kids making a path for us, and were greeted by an elderly gentleman who identified himself in broken English as the "chief" of Balicasag, and clearly more importantly, as an employee of the government as the keeper of the lighthouse. He seemed to know without asking that we were Americans. We gave the cigarettes to the chief, and he distributed some to a few of the other adult males. He asked if we would like to see "his" lighthouse, and we eagerly accepted his invitation. The half mile walk to the lighthouse was on paths traversing well-kept gardens and neat thatched houses. Goats, pigs, and chickens far out-numbered the people. En route, we passed a rustic school building, with the children in the yard being led by their teacher in the song "Mary Had a Little Lamb," in English. We felt sure that news of our arrival preceded us, and that this was being performed for our benefit. Along the way, the chief assured us that, thanks to their hero, General Douglas MacArthur, Americans would always be welcome on his island.

We reached the lighthouse, and proceeded up the interior concrete stairway. Halfway up, the stairs opened to a room, in which sat a desk (his), and on the wall hung a guitar. The chief paused here, and asked if we would like to hear his favorite song. Of course! As we stood there still wet, with our flotation vests around our neck, and our fins and masks in our hands, the chief plucked the chords on his old guitar, and, in the most strong and clear voice sung for us "God Bless America." Tears came to our eyes. What a touching and wonderful moment! This was a proud man. And he made us proud that we were Americans. He gently hung the guitar back on its peg, and we finished our climb to the top of the lighthouse, where the view of the entire island, and the reefs surrounding it was absolutely spectacular. When we walked back into the sea to depart, the children waved and smiled as though they'd just been visited by the Royal Family. Most had bad teeth, as there obviously wasn't a dentist on the island, but their smiles were a mile wide. I'd really like to return there some day....

It's Been a Blast

So many adventures, new friends, unforgettable scenes and scenery, both above and below the water! It's true enough that when one makes a vocation out of an advocation, there is often far less time available to pursue the advocation. Making a business of a sport usually means less time to actually enjoy the sport itself. Most of the time that was indeed the case for us. And, of course, even these two or three times each year when we did get to travel and dive there were responsibilities to be mindful of. Oh, but the fun we've had! I wouldn't trade it for the world!

11

Scuba Students to Remember

In 1987, I "retired" from scuba instruction. At the time, I was involved in a great deal of business travel, and it became difficult to schedule regular classes, and prepare adequately for them. Besides, by then, I had trained over 2,500 new divers, and helped to certify over 100 instructors, who went on to teach thousands of other new divers. Our company's employees were now teaching most of our classes, and, besides, maybe enough was enough. My very last class consisted of a number of good personal friends and neighbors; our daughter; our son-in-law; and our granddaughter (who has since provided us with some great-grandchildren!) It was a special class, for me, and we traveled for our field trip dives to Crystal River, in Florida, where everyone got to dive with the manatees.

The Backward Swimmer

In that 26 years of introducing others to the sport I loved, I worked with a lot of different types of people. Still today, I get calls and letters from folks who were in some of my earliest classes, and, sometimes when I'm in our store, a customer will say "Hi Bill! Remember me? I was in your Columbus YWCA class in 1972!" I've long since confessed to myself that I can't remember the names of all of these folks, but I've learned ways to usually get them to say their name before they realize that I don't. Still, some of their names stick with me, because of something unusual about them, either on the "plus" or "negative" side. Many of my students have gone on to careers such as marine biology, research diving, charter boat operation, resort management, etc. That's always rewarding to think about. Others of them stand out in my memory mostly because they were difficult to work with. Nothing's easy, and every teacher needs some problem students! I had my share.

Many were fun, at least in hind sight. Others were difficult because they didn't really want to be there.

Sometimes an enthusiast will "drag" a reluctant spouse into a class, in hopes of enthusing them, or at least in hopes of creating a built-in dive buddy. Sometimes, these spouses, male and female, have developed into skilled and active divers. More often, however, I have wished that they had less determination to, by golly, "finish that class if it kills me!" It has always been my practice to stick with a student until they can meet the certification requirements, even if they must repeat several courses (at no charge), to do so. Some have required three or four times around. Usually, I admire this kind of determination. There have been times, however, when I sincerely wished that some of these folks hadn't been unwillingly "forced" into taking a class. Still, I've always felt obligated to work with them as long as they were determined to finish.

Don't misunderstand. Learning safe scuba is a snap for a motivated person with average intelligence and at least some sense of confidence in the water. The vast majority of scuba trainees breeze through the course and thoroughly enjoy it.

My own wife, Alice, bless her persistence, was one of those who had some set-backs in her scuba training, but still turned out to be an excellent and enthusiastic diver. I'd been diving about seven years when she first indicated an interest in learning. I had, from the start, promised myself that I wouldn't push her into it. In fact, I'd been diving several years before she even learned to swim. Actually, there really isn't much swimming involved with diving, but one must know the swimming basics, anyhow, and, of course, it is a requirement for certification. Before Alice even thought about learning scuba, she decided that if she and our children were going to be around the water she and they had better learn to swim. She and our two youngest enrolled in a series of swim classes, and became proficient. Now more confident in the water, she next decided to learn scuba. (By now we were grandparents!) However, once enrolled in a scuba class, she had a problem with completing the required underwater breath-hold swim distance. Try, try, and try again, she just couldn't make the 20 yards, despite the help of a number of skilled swim instructors.

She had to drop her first scuba class and try again. (*I wasn't going to be accused of lowering the standards for my own wife!*).

The second time around she still found it impossible to swim far enough under water to meet the requirement. Everyone tried to help her. Finally, I said "Look — let's at least isolate whether it is your arm strokes or your leg kicks that are your weak point, and then we can work on that!" I had her swim a lap using her arms only, and she seemed to move through the water quite nicely. Then, I gave her a kick board to hold with her hands, and told her to do a pool length using only her legs. Thrashing furiously with every leg stroke she knew, she moved backwards! Now, she's only five feet tall, and has short legs, and couldn't be expected to have a really powerful kick, but she clearly did not have a feel for pushing against the water with her feet and legs. All this while, her leg strokes had been working against her arm strokes! No wonder she couldn't make any speed! She worked on her kicks, and eventually solved this problem. Finally ready to proceed with the rest of the course, she then developed some medical problems that prevented her from finishing this course. In the end, she finished her remaining requirements in another instructor's class. She passed with flying colors, and conducted her field trip dives during a shop trip to the Cayman Islands. She has long since developed into a very capable diver, and an excellent underwater photographer. This was a perfect example of persistence paying off!

The Whale Rebels

For several years in the '70s, I taught a scuba course on the campus of Kenyon College, at Gambier, Ohio. The pool there was originally an outdoor pool that had been made into an "indoor" pool by covering it with a greenhouse. Though heated, in the winter months the air inside was frigid, and the water temperature more suited to polar bears. This tended to make our water work move right along, because everybody wanted to get everything done and get into the hot showers! We were given the choice of using as our classroom a room in a building clear across campus, or the men's locker room. We chose the locker room as being more convenient, but, I must admit, this was probably the most odoriferous classroom I've ever used.

Kenyon, though a Baptist school, tended to house a fair number of unorthodox characters. This is the school from which Jonathan Winters was expelled for streaking naked across campus while painted green on a St. Patrick's Day. All of our scuba students there were adequately motivated ("Hey — Dad & Mom are paying for it, and it sounds like fun!"), and most were physically fit and more-or-less of an academic frame of mind. Some of our graduates from there still keep in touch with me. A few, though, were clowns and wise-apples, making for some memorable scenarios.

One Kenyon student, so obese he was nick-named "Whale," was a good and attentive scuba student. He did extremely well with the classroom tests, and seemed adequately skilled in the water, despite his girth. On the last night in the pool, however, he refused to pass through a "tunnel," comprised of two open-ended steel barrels welded together end-to-end, that we had placed on the bottom of the pool. Doing so was part of a sequence of activities designed to test the students' reactions to "adverse situations." The Whale would complete all of the other steps in good order, but when it was time to swim through the "tunnel," he would swim *around* it! We had him repeat the whole exercise over and over, since we were going to insist that everyone complete this sequence exactly as assigned before they would pass the course. Whale wouldn't go through the tunnel! "Are you afraid that your size will cause you to get stuck in the tunnel?" "No." Then why was he being so obstinate? Finally, he explained. Throughout the course, we had often repeated to the students that they should never start any dive they did not feel comfortable undertaking, nor perform anything under water that might lead to undue apprehension. That is, if a diving buddy says "Let's go 80 feet deep." and you don't feel like going to 80 feet, DON'T! "Stay within your confidence limits, and don't expose yourself to undue stress!" Whale applied this concept to our insistence that he pass through that tunnel, and determined that he was being asked to do something that took him outside of his zone of confidence. He felt that he was proving to us his good judgement by exhibiting the cautions we had been drilling into the class. He argued his case well (he has since become an attorney), but we explained that this was a learning experience and an exception to the general rule, and, in the end, got him through that tunnel. Once through, he proceeded to swim back and forth through it, now thoroughly proud of what he had accomplished!

Simulated Emergencies

The exercise the Whale was required to complete was commonly known as "Harassment Night." During the assigned sequence of underwater activities, the instructor would create artificial "emergencies," such as flooding the students' masks, partially shutting off their air supply, entangling them in ropes, etc. The purpose, of course, was to condition the students to respond to each situation in a calm and thoughtful manner. Just as in driving on the highway, a reaction made in PANIC most often leads to a worsening of a bad situation. Cool is cool, in diving. Instructors, of course, can get quite imaginative about creating "simulated emergencies," and, I must admit, some instructors sometimes get carried away. At this point in another course, I had briefed all of the students as to what was going to take place. I emphasized that "Absolutely NOTHING is going to happen that cannot be resolved under water! You will NOT need to surface to fix something! Simply think it out, and a correct level-headed decision can be found to correct any situations that might occur." The students were teamed in pairs, and instructed to "look after" each other during the exercise, and to lend help, if the other buddy got into a "situation" and needed help. And, as usual, I cautioned them against being "defensive" against the instructor, and to try to pretend that he simply wasn't there. The first pair proceeded into the assignments, and were nicely handling the inconveniences I was creating. Gosh, I was good at this!

Then, I completely removed the mask from one of the divers and proceeded to the surface, where I intended to throw it through the air to the other side of the pool. The solution we hope for, in this scenario, is that the buddy still with his mask intact will see that the other diver has lost his, and, guiding his blind buddy, initiate a search to find the lost mask. It was, I felt, a useful exercise. In this instance, however, I reached the surface immediately under the low diving board, and, not realizing this, smashed the mask against the edge of the diving board when I attempted to throw it. The mask's lens shattered, the metal rim was bent completely out of shape and came loose from the mask, and the pieces all fell to the bottom, right before the eyes of the assisting buddy, who already had his buddy by the arm and was searching for the mask, exactly as he should have been. He stared at the pieces, broken glass and all. He began to gather them up, and I could swear I saw a big underwater question mark forming over his head! After all, he had been clearly told that nothing was going to happen that couldn't be taken care of under water! But how the HELL was he supposed to put this mask back together!?

He finally looked up at me, and I've never seen a more questioning expression formed inside a mask. I removed my own mask and dropped it to him, calling to an assistant for another for myself. Luckily, my mask fit his buddy, who cleared the water from it, then stared at the wreckage of his own mask on the bottom, and they resumed the exercise. When they had finished and were back on deck, I decided not to say anything until they first voiced a reaction. When it came, it was a classic: "Gee, Bill, wasn't that a little DRASTIC??" Of course, I bought the guy a new mask....

Students Fight Back

I remember one more incident regarding a "simulated emergencies" class. After reminding the class that they were to totally ignore the instructor during this exercise, no matter what took place, I assigned buddy pairs to begin. A young married couple was the first, and, as usual, I allowed them to complete the first several underwater work stations before I even entered the water. This, I found, usually allows the students to relax a bit before the "emergencies" begin.

And, when the buddy pair is guy/girl, I often created a couple of incidents for the guy first, before starting on the girl. Call me a softie. In this case, as I reached to remove the husband's mask, the tiny little wife knocked me sideways with a violent body block. I shook my finger at her, signaling that she was not supposed to be defensive. She glared at me. I turned her husband's air valve off. Before he had drawn the next breath, she had it turned back on. I headed for her with a coil of nylon rope, intending to get her tangled up. She fled the scene, pulling her husband along with her. When I caught up with them, I reached for her vest inflation button, to see how she would deal with unexpected flotation. She gripped my wrist to keep me from messing with her controls. I looped the end of my rope around her husband's tank valve, intending to drag him away from his buddy, who was going to be temporarily "blind" by my flooding of her mask. She snatched the rope from my hands, and with her other hand kept me from loosening her mask. I gave up. Back on the pool deck, I confronted her. Instead of being repentant about her behavior, she was quite proud of having proven to be a GOOD and RELIABLE dive buddy. In no uncertain terms, she explained that my training had clearly stressed the importance of looking out for one's partner, regardless of the circumstances. As far as she was concerned, this over-rode my instructions to "ignore" the perpetrator of the problems they had just faced. Privately, I conceded her the point, and, personally, I felt that this woman would, indeed, make a darn good buddy!

Coal Buckets

I don't know who started the phrase, but my assistants would sometimes privately refer to a student behaving in a dense manner as a "coal bucket," (as in, dumber than a...), or a "dipstick" or a "twitch." Not flattering for sure, and it wasn't our habit to ridicule anyone, anytime, but, by golly, we would now and then have a student who was pretty hard to get through to. For example, one such fellow was having such a tough time understanding anything or getting anything done that I began to wonder if he was mentally competent. Still, I was determined to do whatever it took to get him through, if he had the will to keep trying. He did have a decent job, so I assumed he must be at least somewhat together. Then, one night during a break in the pool work, I stepped into the public rest-room to relieve myself. Here, at one of the wall-mounted urinals, was this same red-headed guy using the urinal *with his BACK to it and his butt up against it*, as if that was the most natural thing in the world! Now, this fellow was probably in his early twenties. Could it be that in that many years he had never understood how to use a urinal? Or, did he really know, and just forgot? Surely, he had seen others doing it the right way — did he think *they* had it wrong? From the bland smile on his face as I walked by, he very obviously wasn't doing it as a joke on me, as he'd have had no way to know I was going to walk in.

Now, I don't really like to think of myself as having a narrow mind, but this act convinced me that we were likely to continue to have problems working with this Mortimer Snerd. Sure enough, before we had to throw in the towel on him he stopped showing up for class, probably going on to bigger and better things. I wonder what?

Say What?

One of my favorite teaching innovations was a portable underwater speaker system, with which I, on the surface, could talk to students under water. The system was comprised of a speaker suspended into the water and an amplifier and microphone on the surface. The speaker incidentally also served as an underwater microphone, and one could, on the surface, hear the student's breathing and clinks and clunks as they moved around on the bottom. Late in the course, part of a pool session involved the students, in pairs, progressing through a series of "work stations" on the bottom of the pool. Each of the teams had a number, as did each of the work stations. These stations involved such as assembling and disassembling metal puzzles, exchanging masks, buddy-breathing, and so forth. I, on the deck, would randomly direct the teams from one station to another, thusly, I thought, teaching them to listen to and follow instructions. Once I could see that each pair was finished with their station, I'd call out for "team number four to proceed to station number five," etc. Though I usually liked to be in the water with my students, for this exercise I rather enjoyed playing the "boss" with my top-side mike.

One night, in the middle of this exercise, the clear message boomed from the amplifier: "WHAT STATION SHOULD TEAM NUMBER TWO BE WORKING ON?" What? Nobody had ever "talked back" to me during this exercise! Amazingly, a very clever student, not having understood my directions, had approached the underwater speaker. Cupping his hands to form an air pocket to speak into, he removed his mouthpiece and enunciated his question. It came through crystal clear. The student had never seen this done before — he simply figured it out for himself. I answered his question and the class finished the exercise and exited the pool. I congratulated the young man on his ingenuity. He explained that he had been watching our staff set up the speaker system before the class started, and simply figured out by himself how it worked. That I could understand, but forming words under water is impossible without an air space to speak into. I asked how he had worked this out, and he answered that cupping his hands had simply occurred to him. I've always admired the skill of innovation, and this guy was sharp. After his course, by the way, he went on to become an expert and enthusiastic diver, and an outstanding underwater photographer.

Domenic the Pest

Though we also taught private classes to couples, families, or small special groups, the majority of our classes were groups of 10-15 strangers. Many, many long-lasting friendships originated from these classes, and even several marriages. Some classes even have annual reunions, counting their scuba course as being a turning point in their lives. Now and then, however, we would get a real "loner" in a class. Not only did these kinds not care to become acquainted with their fellow students, but some of them clearly wanted all of the instructor's attention for themselves.

Domenic Whiner was one of these. That's not his real name, but is most descriptive of his attitude. I should have been aware of a probable problem when he signed up for class, because he told me he had been "kicked out" of a competitor's class. I like challenges, however, and felt sure I could work him through the course.

Eventually, I did. That is, after going through a couple of courses he finally completed all of the requirements and passed all of the tests. Still, I found myself hoping that if he actually got into diving he wouldn't tell anyone who his instructor had been.... Domenic was like a leach, throughout all of his classes. To him, the other students didn't exist, and he constantly insisted that I work one-on-one with him. I couldn't short-change the other students, and explained that to him. He cried. He whined. He'd been neglected at home, he said, and really wanted to learn this sport, so he could have something exciting to do, and could "meet other people." Later, I learned that he had been such a pest at home that his own parents left him the house and moved to Florida to get rid of him. As to "meeting other people," these classes were an excellent opportunity to do so, yet he totally alienated all the other students. His whole diving career was like that. Clubs didn't want him to join. Shops didn't want him to sign in on their trips. He wanted special price concessions on every piece of equipment he bought. In short, he was a pain in the ass. Everybody needs one, I guess. Domenic was mine for years, as he "adopted" me.

My office at the shop was out of sight, and I instructed my staff to tell Domenic I wasn't in, when he stopped by. "But his car's out front!" he'd scream, and I'd usually end up coming down to deal with him. In the water, on a dive, he would invariably get in some sort of trouble, requiring someone else to help him and generally messing up everyone else's enjoyment. He was a poor diver, in every respect. Yet, I had certified him. If it had been in my power, I would have committed him, instead of certifying him, but he passed the course, and had my card. I somehow felt obligated to keep track of him. He quit diving, at least twice, and sold his gear each time, but then got back into it. Part of the pleasure of my "semi-retirement" is that I don't have to deal with Domenic anymore....

Lieutenant Rooster

One of the first working recompression chambers in the midwest was installed at the University of Michigan. Though also used for medical research and experimentation, this chamber was basically funded for the treatment of accidental diving maladies, such as decompression illness (the "bends"), etc.

Often, this chamber was used for experimentation with improved treatment procedures, simulations of deeper dives, and various other diving-related studies. We also often took groups of instructors to visit the chamber and sometimes experience simulated dives and "treatments."

A contingent from the U.S. Navy once used this chamber for tests regarding the effects of nitrogen narcosis. Also known as the "rapture of the deep," narcosis is the effect of breathing air under pressure. The greater the pressure, or depth, the greater the "narcotic" effect of the nitrogen in the breathing air. This effect is actually taking place, to a very mild degree, even at shallow depths, but presents no particular hazard. At a depth where an observant diver could actually begin to notice the effects, perhaps around 60 to 90 feet, for most of us, the symptoms could be best described as a modest "euphoria," or sense of well-being. Much deeper, and the symptoms resemble intoxication, with the diver getting "drunker" (and dumber) the deeper he or she goes. Clearly, being this deep is not a place where one wants to be stupid! This is one of the main reasons why sport divers should limit their diving to depths of 80 feet, or less, in my opinion.

When one is experiencing these narcotic effects at a certain depth, they do not increase the longer the time spent there. In other words, when you reach a depth of 100 feet, you are as "drunk" as you are going to get. Ascend back up to a shallower depth, and the effects diminish at once. And, unlike alcohol intoxication, no after-dive "hang-over" results. If you didn't do something dumb enough to endanger yourself or your buddy while you were at the deeper depth, no ill effects will transpire. There is, by the way, absolutely no correlation between one's tolerance for alcohol and the effects of nitrogen narcosis.

Back to the Navy's experiments in the Ann Arbor chamber. Since military divers often may have need to dive beyond the sport diver's safe range, it was this group's intent to study the narcotic effects of breathing air at depths of 150 to 300 feet. In the chamber, they could simulate these dives by simply pressurizing the sealed air chamber, with the men inside, to a pressure equivalent to whatever depth they chose. Thusly the men could experience deeper "dives" without even getting wet, and with all types of safety control systems in place.

Lieutenant Salsbury was in charge of the group, and decided to participate personally in each of the tests. A strict and sober disciplinarian, he actually thought that his own rank and rock-steady personality would serve as a control benchmark against which the intoxication levels of the lesser enlisted men could be measured. He had many times over proven that he could "hold his liquor," and had never been drunk. In fact, he was as "sober" a red neck as one would want to be around, with no sense of humor whatsoever. These tests were to be "all business," and he warned his men against any joking around.

He and two of his men were locked into the chamber for the first experiment. The air began to hiss into the compartment where they were seated, and the "depth gauge" started to climb. A movie camera was trained on the three men through a glass port, and a microphone was attached to each. As the "depth" was increased to 80 feet, 100 feet, 120 feet, the two enlisted men were seen to physically relax, and their facial expressions gradually changed to very "contented." Still, they made no wisecracks, as the Lieutenant had clearly described to them how "professional" military men are to conduct themselves. The lieutenant himself continued to sit ram-rod stiff, with a very stern and solemn expression. "So this is what officers are made of," one of the outside observers said to the others. The depth continued down, and the enlisted men's expressions turned to wide grins. Not the lieutenant, who remained rock steady. Exactly when the depth gauge read 180 feet, the lieutenant stood to his feet, started flapping his arms as if they were wings, and with a gloriously wild expression started crowing like a rooster! "Cocka-doodle-doooo!" "Cocka-doodle-doodle-doooo!" Scratching one shoe on the floor, then the other, he threw his head back and crowed ever louder, "COCKA-DOODLE-DOOOOOO!" The enlisted men looked up at him in awe, then burst out laughing. The chamber operator immediately started bringing the pressure back to normal. At a depth of about 150 feet, the lieutenant suddenly sat back down and immediately resumed his previous look of stark soberness. His men, likewise, got themselves stifled before he noticed their hilarity.

The chamber was soon brought back to normal pressure, opened, and the men exited. "Good work, men," said the lieutenant. His men couldn't bring themselves to relate what they had witnessed. One of the civilian operators, however, asked the lieutenant if he realized what he had done. "Certainly," he said, "I observed my two men having mild euphoric effects, but nothing at all major." "Next time, we'll go a little deeper." The operator told him that, at a depth of 180 feet, he had crowed like a rooster and flapped his arms. "Nonsense!" As is rather typical, he had absolutely no memory of what he had done. The next day they showed him the film. Later that day he and his team departed, explaining they had been called back to their base for an emergency, and would have to cut the chamber tests short. The lieutenant said that since they had only made the one dive, the tests could not be considered "official," and that he, personally, would take the film with him, until the tests could be resumed at a later date. They let him have it (but not before they made a copy!).

Teeny-Bopper Divers

In the early '80s, a local field study group adopted the practice of taking groups of high-schoolers on field trips to the Bahamas Islands, and allowing them to scuba dive there, with absolutely no formal instruction. Once aware of this, I lobbied strenuously against the practice, and eventually the rule became that the kids would have to complete a full scuba class prior to going on one of these field trips and diving. Though it honestly was not my original intent, our facility was called on to teach some of these classes, and I was personally in charge of some of them. Many of the youngsters were eager to learn and proved quite proficient. In each class, however, were several who were obviously only present because they "had" to be. Mom and Dad were paying for the course, they had no real interest in diving, and they had heard that these "field study trips" to a tropical island were great "boy-girl" adventures, and much more fun than staying back at school. Getting through this class I was trying to teach was, to them, just basically a bore. "Do we *have* to do this?" "Do we *have* to do that?" "Can't you just give me this dumb card, so I can take the trip? I promise I won't dive while I'm there!" Frankly, I found it difficult to have patience with these types, and many of these classes were as near to being "work" as any other classes I've ever taught. Still, at least, this group is no longer taking untrained teen-agers on salt-water scuba dives!

Double Exposure

We were working with a large class at the local Jewish Community Center. This night, one of the student's assignments was to enter the water, proceed to the bottom, and, while kneeling, remove their scuba unit and mask. They were to remove the scuba unit by pulling it up and over their head, lay the unit on the bottom in front of them, with their mask, then re-don both, and return to the surface.

Because of the size of the class, I was having two of them perform this exercise at once, with me on the bottom, wearing scuba, to observe and assist them. My assistant, on the deck, was preparing two at a time to enter the water and proceed to my position. Things were moving along quite well. When the third pair came to the bottom, both were attractive shapely young gals wearing two piece swim suits. Both had been doing quite well in class, taking all of the previous exercises in stride and performing everything with calm and polish. Tonight, as I watched the first girl reach over her head for her tank, both of her breasts popped out of her swim suit halter. By the time I turned to look at the other girl, the very same thing had happened to her! Now, nipples were staring at me, from both sides of the pool! Should I go to their assistance? Exactly what could I do? I stayed put. Both girls coolly finished their exercise, re-secured their scuba unit, donned and cleared their mask, and calmly stuffed their breasts back into their suit, as if nothing had happened. Back on deck, after all of the other students had completed the exercise, nothing was said about the incident, and the girls seemed eager for their next assignments. Nobody else had witnessed the scene, and I was quite sure that neither young lady had seen the other's predicament. Perhaps both thought that I was looking the other way when they lost their whatchama-call-its. I never told. (Till now...)

Bare Facts

This incident didn't occur during a scuba course, but was somewhat related. Our new building had a full-sized pool, and this required quite an investment. Scuba classes alone didn't keep the pool busy or pay the bills, so we also had open swim periods, taught swimming classes, and rented the pool out for private parties. The first winter, we were approached by a large nudist club who wanted to rent the pool on Saturday nights, twice a month. We sure could use the rental fees, so we agreed to try it, on a trial basis. They assured us that they were just "regular" well-behaved people, and would be no trouble at all. Our attorney advised us that no legal problems would be involved, if the nudist swims were truly private affairs. The law, however, still demanded that we provide a certified life guard. Still not convinced that a bunch of naked people might not present certain security concerns, I felt that I, personally, should serve this duty for the first few sessions. Unselfish, you say? Whatever...

In fact, I was quite pleased to observe that they were, indeed, just like any other group of folks enjoying a swim party. In fact, several whole families were involved, including children of all ages. And, they never even suggested that I, the life guard, should remove my own suit. After the first half hour or so my apprehensions disappeared (though my suit didn't). I must admit, however, as I policed the mens shower/locker rooms, which both sexes shared equally, I was far more ill at ease there than in the pool. Somehow, seeing males and females undressing together seemed more embarrassing than seeing them all totally naked in the pool.

The first night, a young couple from out of town approached me with an idea. They alone, among the club members, were scuba divers. They asked if they could bring their equipment to the next nude swim party, and provide a diving demonstration for the rest of the club. "Certainly," I said, and they did. That night, the young wife was assembling her scuba unit on the deck when she experienced a problem connecting the regulator to the tank. She asked for my help, which I readily provided. As I knelt at the tank, she stood immediately behind it, with her crotch at my eye level. Now this scenario alone I probably could have handled. But, sometime since the last swim party this attractive young lady had shaven off all of her pubic hair. I moved around the tank to get this picture out of my line of sight. She moved opposite me. Not at all provocatively, but simply to help steady the tank. I finished the job as quickly as possible, and removed myself from the area. Nobody has ever accused me of being a prude, and, under normal circumstances, I might have found the scene titillating. Or, perhaps, I found it too titillating. In any case, I'm quite sure that I blushed dreadfully, and could think of nothing else than to get elsewhere, and quickly. The scene, however, has obviously stayed with me.

Slow 'em Down, Ned

It was our practice to conduct our student's field trip dives in as clear water as could be found, though this usually involved nearly a 200 mile round trip for most. Those classes conducted in the late fall and winter months would, of course, have their field trip dives postponed until spring. From time to time, an entire off-season class would instead opt to travel to Florida or an island resort for their field trip dives, in the winter months. Seldom did the paid instructors argue with this arrangement, but it was somewhat of an exception.

Since the instructor can't personally make the required number of dives with each student individually, trained and certified divemasters are often employed. In most cases, these assistants work on a volunteer basis, or nearly so, because they enjoy introducing new divers to "their" world.

Two of the most enthusiastic assistants who worked with me were my good friends Jack Porter and Ned Neibarger. Jack had been my main dive buddy after we moved to Newark, and Ned was in one of my classes a few years later. Both were somewhat "middle-aged" when they started diving. When Ned returned to shore from his own very first open water quarry dive, he was absolutely in AWE of the experience. He rushed to me and exclaimed "WOW! What an experience! The whole time I was under water, I had the distinct feeling that my whole self was inside that mask, and looking out through it as though it was a picture window!" He never lost his enthusiasm and, once he had logged a sufficient number of dives, he was eager to be assigned as a dive buddy with new students, especially on their first dives.

Ned had a very relaxed diving style, and an extremely low air consumption rate. A single tank of air would last him through at least three student dives. In fact, if the agenda called for the assistants to take the new divers down for, say, 30 minutes, Ned and his buddy would usually stay for 45 or more, sometimes holding up the day's agenda. Unlike a lot of divemasters who persisted in showing the students as much of the bottom as possible, Ned would slow his student down and show him or her how to really take the time to see and observe the underwater life. Slow is good, under water, and Ned was good at demonstrating the benefits of "looking closer."

Years later, far more former students would fondly remember their dives with Ned than with any one other assistant. Later, Ned's younger son, Don, would also become a valued helper with our classes, and Ned's wife, Reta, though not a diver, would often be present to help with shore duties. Sadly, Ned succumbed to liver cancer a few years ago, and Reta was later killed in an auto crash. We miss them dearly.

The Hot Water Trick

Field trip dive excursions in the early spring were usually an adventure in learning what *cold* really means. We would usually start our dives in mid-April. While the air temperature was warm enough, the lakes in Ohio haven't really warmed up much until about early June. Water temperatures of 55°F to 60°F are common in the spring. That might not sound all that cold, but that is the usual temperature of water from a cold water faucet. It is bone-chilling. The diver's wet suit, hood, gloves, and boots provide adequate protection for a span of time, but stay long enough and the body will be thoroughly chilled. And, of course, when first submerging into this water, a layer of this cold water enters the suit with the diver. In only a couple of minutes, that pint or two is warmed to body temperature, but those two minutes can be a real shocker to the un-initiated! We once had a student show up for his second field trip date with a two-quart thermos of hot water. Just before entering the cold water he poured this water down the neck of his jacket, thusly avoiding the icy jolt when he submerged. Good thinking! In fact, some others of us thereafter adopted this neat trick!

12

Fond Memories and Good Friends

In this book, I'm hopefully not looking back at my entire diving career. With luck, I've only yet experienced perhaps two-thirds of it. Though I'm not out diving the colder quarries and lakes every weekend any more, I still look forward to each new trip to warm, clear sea water with as much enthusiasm as I always did. In fact, I and my good friend, Jack Porter (age 66), have just returned from two week diving excursion to Micronesia, diving the WWII Japanese shipwrecks of Truk Lagoon and then on to Palau.

In this *first* 40 years, nonetheless, I've enjoyed a wealth of unique experiences, and have made friends with hundreds of truly fascinating people. Diving has, indeed, opened up three new worlds for me. First, the underwater world, which so many people live an entire lifetime without experiencing. Second, the world of travel, and the opportunities to have seen so much of this little planet we occupy. And, thirdly, the world of people who share an enthusiasm for life and living it.

Those around me have countless times heard me express my appreciation of having been able to make a career in this unique and exciting field. Certainly, through the years, the building of a new business has involved a good deal of anxiety, plenty of problems, money concerns, gazillions of hours, and, undeniably, a certain amount of stress. Still, all this while, how splendid to have been surrounded, day in and day out, by employees and customers who derive enjoyment from what we're doing. I've often reflected how different it would be to instead be involved in a business that depended on someone else's misfortune. Picture the operation of an automobile body repair business, for example, in which every customer to be dealt with has suffered some bad luck! Or the roofing business, in which every client is distressed by the need to require your services, and not likely to get all that much thrill or enjoyment from your finished work. Our business, on the other hand, has not been just the business of teaching diving and selling diving equipment, but, in truth, has really served the function of helping folks find *recreation*. Customers who come to us are seeking enjoyment, adventure, and excitement. The vast majority of them develop into real enthusiasts, and what could be more fun than dealing with people who are really "in" to what they're doing? What a lucky guy I've been!

Just how much has "luck" played a part in all of this? Plenty! Yes, I've received some awards and recognition, and our business has prospered and grown. But the biggest element in all of this, by far, has been the "right place at the right time" factor. Diving was just *starting* to become popular when I stumbled onto the scene. Scuba instructors were in great demand, just when I lucked into that opportunity. Equipment was difficult to buy, when I blundered onto a way to capitalize on that. Manufacturers were just then establishing more outlets, when I happened to be positioned to benefit from that. Few people in the country knew much about air compressors, just when my basic experience gave me an edge. And, above and beyond all of this fortuitous timing, good fortune brought into my life the most marvelous array of talented and dedicated people, to help make all of the enterprises work. And above even this, I've been blessed with a wife and family that have been behind me 100% the whole while! Yes, indeed, I've been a lucky guy! And, somehow, my Lord has seen fit to get me through a couple of pretty serious cancer operations. Who couldn't take *that* as a sign that He still has something for me to do?

Along the way, I've met and maintained contact with a lot of lively characters, all around the world. Many were folks who had dedicated their lives to the advancement of the sport. Others had in many cases "forsaken civilization" and established sport diving operations in remote areas where a true "pioneer spirit" was required. And, some had devoted their lives to a nature study of specific aquatic species and/or habitats. Individualists all, yet in each and every case these were people who were eager to share their interests, findings, and enthusiasm with any fellow diver of a like mind. I have been more than fortunate to have enjoyed contact with so many of these lively people, and to have been instrumental in exposing so many of my fellow enthusiasts to them.

Fish Spotting

I've mentioned before that a large percentage of divers who adopt diving as a life-sport get intrigued by underwater photography. Many others likewise find great satisfaction at "cataloging" the various species of underwater life they observe. Some combine the two interests, and catalog their observations with photographs. Lots of non-divers find interest in bird-watching. Some even travel around the world to spot and identify exotic bird species. Many of the real bird enthusiasts maintain a "life log" of the birds they've spotted, always striving to fill in as many blanks as possible, while at the same time being aware that they will likely never get to see them all. Several years ago I promoted the concept of the "life log" for divers, wherein a traveling diver, with identification books, could strive to find, identify, date, and record as many as possible of the various forms of a specific type of underwater life. I've been most gratified to see many divers pick up on this objective. Some target fish exclusively, while others focus on coral varieties, sponges, shells, anenomes, etc.

Underwater Undergarment

One of my own special interests in this respect, has been the filming of nudibranchs. These are, in effect, snails without shells, and most are quite small, being typically only an inch or two long. Hundreds of varieties exist, and many have developed elaborate spectacularly-colored shapes, resembling exotic live moving flowers. They are not, however, all that plentiful or easy to locate. Whenever we have a group diving in tropical waters, I always request that everyone keeps an eye out for a nudibranch, and to signal me if they find one. In waters where they exist, I usually keep a close-up lens on my camera. One of the largest species is commonly known as the Spanish Dancer, and can exceed 8 inches. Somewhat rare, and found only in certain waters, the Spanish Dancer is brilliantly red, and like other nudibranchs, doesn't swim free. However, if one is removed from the coral and released in mid-water, it undulates in the most remarkable and graceful manner until it slowly returns to the bottom.

World-famous underwater cinematographer Stan Waterman has related a Spanish Dancer "episode" he once experienced. Diving with a group in the Red Sea, he had requested all of the other divers to watch for a Spanish Dancer he could film. None were found all week. On a night dive on the last day of the trip, however, a diver excitedly came over to Stan and pulled him to the other side of a coral head, where a group of divers were spotlighting with their underwater lights a splendid scarlet Spanish Dancer "dancing" in mid-water. Stan rushed in, turned on his movie lights, and started filming. After a couple of minutes, however, he noticed through the close-up lens what looked like stitches on the creature. Looking closer, he realized he had been duped. His "Spanish Dancer," undulating so beautifully, turned out to be a pair of the cook's red panties which the other divers had mischievously positioned, billowing and tumbling gracefully in the gentle current! His revenge remains un-recorded....

Local Divers Land Giant Carp

Back in the '60s, some of my friends in the local dive club got into the newspaper with a little hoax. Diving at the downstream side of a small dam on the local river, they discovered that the 4-foot-deep pool at the base of the dam was usually filled with many fish, most of which were carp. In short order, they learned that they could actually catch some of these fish, by hand. One day, they hauled out a very large carp, perhaps a ten or 12 pounder.

Local club member gets faked "giant carp" photo in local paper.

Then, through trick photography, against the backdrop of the dam, they produced a photograph of themselves and the carp, with the carp appearing to be fully six feet long. They submitted this photo to the newspaper, along with a fabricated story of having stalked this monster "phantom" carp for several months, before finally wrestling it to shore. The photograph and story was duly published, with the photo caption reading "Local divers capture 100 pound carp!" Once this was published, they could hardly confess the truth, so they continued to embellish the whole story when folks asked them about it. "What a fight!" "Yep, he was a *monster*!" To the best of my knowledge, the truth was never told. (Till now...)

Charlie's Missing!

In the indoor inter-club competitive events sponsored by our State Diving Council, underwater "hockey" was far and away the most physical. Five-member teams, with only mask, fins, snorkel, and a hand-held short wooden wishbone-shaped "stick" would line up on each side of the swimming pool. A goal was positioned at the bottom of the pool wall on each side. The lead puck was placed in the center of the pool. At the sound of the starting whistle, the ten contestants would race for the puck and attempt to push it into the opponent's goal. The scoring of each point often raged for several minutes of frenzied elbowing, blocking, and frantic trips to the surface for more air. Though most wore gloves, knuckles, elbows, and knees suffered plenty of nicks, especially if the pool bottom was at all rough, and the water sometimes got tinged pink with the warriors' blood.

Through the years, this underwater hockey developed into the major event in the winter games. Teams practiced for weeks in advance, and were screened down to the fastest, strongest, most blood-thirsty guys in each club. Our own club, for years, held the underwater hockey championship, and we were proud of it. Other clubs set their sights on knocking us off. I held the distinction of being the best at getting to the puck first, which usually provided a good advantage. Short, muscular little Charlie White (yes, the same guy that blew up that dinghy), proved to be our best goalie.

After one especially lengthy and hotly contested battle with the second place team, our team returned to our side of the pool after scoring the first point. We were exhausted. Red-faced, huffing and puffing, we hardly had breath to congratulate each other. Fully a minute after we had reached the deck, one of the guys said "Hey! Where's Charlie?" We looked to see if he was sitting on the other side of the pool, perhaps having surfaced with the other team. Charlie wasn't there. We looked into the pool. Sure enough, there sat Charlie, butt on the bottom, back against the wall! I dived to him immediately, and was struck by the expression of contentment on his face. He looked for all the world like a happy guy just taking a rest! I hauled him to the surface and the other guys drug him out onto the deck. As we prepared to administer mouth-to-mouth resuscitation, he came around. Absolutely no worse for the experience, he was only eager to learn "Who made the goal?" What a guy!

Pop Goes the Ear Drum

During an underwater hockey practice session one evening, another team member accidentally gave me a sharp blow on the ear with his knee. I knew at once that my ear drum was ruptured. In cold water, this can bring on an instant case of severe vertigo, but the warm pool water only gave me some mild dizzies. I went to the shallow end of the pool, ducked my head under water, held my nose, and forced air into my inner ear. Sure enough, a stream of bubbles flowed out of my ear.

A ruptured ear drum can keep one from diving for a number of weeks, so I was anxious to get this properly taken care of. However, it was now rather late in the evening, "emergency clinics" didn't yet exist, and I didn't want to bother our family doctor. I would go to his office first thing in the morning. Still, inner ear infections can get started within hours, and can be really bad news, so I determined to prevent this by pouring some rubbing alcohol into the ear. The instant I tilted my head and did this, the alcohol ran through the ear drum into my inner ear, and on into my nasal passages. WOW! Talk about pain! It felt as though a red hot poker had been pushed eight inches into my ear. Hopping around yelping in pain, I could barely see straight for several minutes. Whoopee-do! Agony! After what seemed like ten minutes, the pain finally subsided, but I knew I had made a mistake. The next morning, as Doc peered into my ear, he said "Well, you have no infection, but it certainly is RED in there! I bet you put alcohol in your ear, didn't you?" I abashedly answered "Uh-huh." He said "I bet you won't ever do that again, will you?" "Never ever ever!" I answered. He grinned. Doctors are sooo smart. The rupture healed nicely within a couple of weeks.

What Group Are You In?

Remember Pat Stover, the rest room photographer? On a trip to Belize, Pat was still a fairly new diver. Not a scatterbrain, and certainly not a "coal bucket," Pat was nevertheless sometimes "discombobulated." The resident divemaster on our live-aboard boat was very conscientious about checking divers into and out of the water, and he insisted that we place close attention to our dive tables.

What we refer to as the "dive tables" are the decompression limit tables with which a diver works out his time and depth limits for each dive, to avoid exposure to the bends. When planning a dive within twelve hours of a previous dive, the diver must consider that some residual absorbed nitrogen effects still persist in the body from the previous dive, and this will affect the computation of the limits for the following dive. Consulting the "tables" makes this computation quite easy. At the time the diver exits the water from the first dive, he or she is considered to have an excess nitrogen level of a certain amount, designated by a letter of the alphabet. For example, a diver having been to a depth of 60 feet for a total time of 30 minutes would be said to be in "Repetitive Dive Group F." The longer this diver is out of the water, the more of the excess Nitrogen is gassed off from the body, and he will progress back up the alphabet to E, D, etc., until, at the end of twelve hours, his nitrogen level is completely back to normal. Thus, the length of the surface interval between dives is a major factor in this consideration.

On this trip, it was expected that each diver would work out his or her own time/depth limitations, by using the tables, and conduct their dives accordingly. After the first dive of each day, the divemaster would check with each diver preparing for their second dive, to make sure they had completed this computation. Just before they entered the water, he would record the time of day, the names of the buddies, and ask each diver "What group are you in?," meaning which alphabet designation.

The first day, I was standing by as the divemaster went through this routine. "Group D," one pair would call out, then enter the water. "Group C," the next couple might say. Then Pat and her buddy came to the entry platform, and the divemaster called out "What group are you in, Pat?" She answered, as she stepped off the ramp into the water, "McBride's group!" clearly not having heard any of the routine that had been going on. Her buddy, and everyone else within earshot, completely broke up. When she surfaced, after the dive, she quietly asked the divemaster, "Why did you ask me that, just before I entered? Everyone on this boat is in the same group, McBride's!" This was Pat at her Pat-est, and many of the folks on that trip still today greet her with "What group you in, Pat?"

Snort That Water

Another of our indoor competitive events was the simple art of mask clearing. The competitors' masks would be placed on the bottom of the pool. At the sound of the starting pistol, all would dive to their masks, don them, clear them of water while still under water, come to the surface, slap the deck, and face the judge, one of which was assigned to each contestant. The judge's job was to time the competitor with a stop watch, and to see that the mask was completely clear of water. Much more than about a half inch of water remaining in the mask would disqualify that contestant. Amazingly, the winning time was often down in the range of only about four and a half seconds!

Now, with practice, it's quite easy to get the water out of a mask while under water. Attempting to do it in a big hurry, however, often results in an incomplete job. Many of the contestants would get disqualified when they surfaced with a bit too much water still in the mask. One fellow tended to win this event nearly every time. He was not only fast, but his mask was invariably absolutely dry inside when he surfaced. He never got disqualified.

Eventually, we asked him his secret. "Simple,," he said, "I put the mask on, clear it in the conventional manner on my way back to the surface, then, just before I break the surface I suck the water remaining in the mask UP MY NOSE!" As impossible as that sounded, we all gave it a try. For us, it *was* impossible! We'd snort, cough, gag, and give it up. Obviously, this guy's nasal passages were put together in a different way than ours!

The Maytag

For several years, each spring, divers from all over the midwest would convene at Ohiopyle, Pennsylvania, to "Run the Yock." The Youghiogheny River, for most of the year, was a scenic river running through the mountains of southwestern PA. Each spring, however, the gates of the dam just above Ohiopyle would be opened, to lower the level of Youghiogheny Lake. For several days, the river would become a raging boulder-strewn flood, with multiple rapids probably qualifying for about a rating of "5" on the kayakers' scale of hazard. An "insider" who worked at the dam would alert us as to exactly when the gates were to be opened, and we would spread the word, and name the date. The Ohio council adopted the practice of organizing the run, enforcing safety practices, and briefing the participants.

The water, naturally, was quite cold, so full wet suits were de rigueur. Usually, wearing a wet suit with no weight belt provides so much buoyancy that it is impossible to submerge. The "Yock," at full flood stage, however, provided many back flows, whirlpools, and undercurrents so powerful that the wet suit alone couldn't be relied on for flotation. Each river runner, therefore, was equipped with an inflated heavy-duty inner-tube, most of which were of truck size. These tubes were to be tied firmly to the person, in a manner so as to not allow separation.

The designated exit point was about five miles downstream, and was one of only two places at which it was possible to walk out of the river valley. The other was only about 300 yards downstream of the starting point. In other words, once in the river, and past that initial exit point, the runners were committed for five miles of total aquatic turmoil.

My first visit to the event was for the purpose of filming it from shore. In later years, I would run it myself, with an enclosed movie camera to get some "first hand" footage. This proved to be a real adventure in chaos, but I managed to get some memorable action scenes! This first year, I determined to place myself just downstream of the starting point, at the initial "escape" site. From this vantage point, I could film the tubers entering the water, and also had a great view of the first fierce whirlpool the tubers would be sucked into. This was nick-named "The Maytag," and was between my position and the starting point.

First, I filmed the group of diver/tubers being briefed at the starting point. This day, there were over 110, and a more eager and confident group couldn't be imagined. All were equipped with a wide array of garb, costumes, and inner tube arrangements. Several had arranged for imaginative ways to identify their club's members, such as painted hoods, colored tubes, club buoys, etc. The river roaring behind them didn't seem to intimidate anyone, though this would be the first time in the river for at least half of them. Bravado and high spirits seemed the rule of the day! All were anxious to get the briefing over with and hit that river! Since all couldn't enter at once, a drawing determined which clubs would go first.

I moved to the escape point and started to film, as the divers began to enter the river, club by club. My film shows groups of happy, laughing, waving divers sliding down the smooth fast water heading for the "Maytag." Then, one by one, in a fast "slurp," they are sucked into the maelstrom of the whirling waters, and disappear from view. Then, one by one, fully a half minute later, they are coughed up out of the downstream side of this gigantic "washing machine," thoroughly tumbled. Colorful hats are gone, even some hoods are missing, and many of the spluttering divers are frantically pulling themselves back to their tethered inner tube. It is exactly at this point that many of them come to realize that this was only the first experience of five more miles of such violent pandemonium. As I zoom in on their faces, as they surface, one can clearly see the decision written on many of their faces that discretion is the better part of valor, and that exiting the river at the first possibility is probably a very judicious thing to do. Indeed, fully a third of the participants, so eager only a few minutes before, swim vigorously to shore at the position from where I am filming. I film them as they wade to shore. The scene would provide an excellent illustration in a dictionary for the word ABASHED. These are humbled people, indeed! Of course, several claim "cramps," "ear problems," "the flu," or a number of other legitimatizing reasons why they couldn't continue, and we take care not to tease anyone about their abbreviated run. After running the river several times myself, in later years, I actually came to admire these folks for having enough guts to quit! In fact, a few years later an individual tubing the river independently was drowned, and we thereafter discontinued the council river run.

The Tinkerer

Back in the mid-'70s, long after good diving equipment was readily available, a young fellow we'll call Tim Burns joined our club. He was welcomed graciously, as were all new members. However, on the first club dive Tim showed up for he pulled out of his car trunk some of the weirdest gear we had ever seen. All homemade, it was comprised of military surplus pieces and parts, and his tank was a spherical globe of fiber glass. Later, we also learned that his tank was filled with air from a compressor he had also fabricated from spare parts.

Tim's equipment seemed to work, more or less, and one certainly had to admire his ingenuity, but everything in our diving manuals warned to avoid the use of homemade equipment. Nobody wanted to buddy with Tim, with his Rube Goldberg inventions, and he was told that it shouldn't be used. "Show me where that rule is written!" he would say, and no such written regulation could be produced. At club meetings, he would bring in his latest innovation and expound on it. The guy was obviously a craftsman, but his theories were out of touch with reality. Certainly, creativity shouldn't be dampened. After all, that's how the world progresses. But Tim's main goal was not so much to advance the technology, but to save himself money. And, many of his contraptions clearly were pushing the boundaries of safety. More than any other error, Tim seemed determined to pressurize many of his components to pressures far greater than they were designed to withstand. "I know it's not rated to that pressure," he would say, "but I've already pressurized it to twice that pressure, and nothing happened!" We wondered how many members of his family were around when he conducted such "tests."

Gradually, Tim became a thorn in the side of the club. Discussions of his persistence in using his "inventions" during the club's pool practice sessions disrupted many meetings. Eventually, the matter was resolved by modifying the club's written by-laws to simply prohibit the use of "experimental" equipment during club activities. This more than rankled Tim, of course, and he left the club. In fact, he left diving. The last we heard of him he was working on a personal one-man airplane. I suppose he built it with aluminum cans and tin foil.

Getting High In the Pool

This anecdote involves my father. A few years after our new building was opened it was time to re-paint the indoor pool. We decided to do it ourselves. (Well, actually, at the time, we couldn't have afforded to hire professionals to do it.) First, the pool had to be completely drained, and allowed to dry. Then, we scrubbed and sandpapered the old paint.

When we were ready to start painting, it happened that Dad was visiting us from out of town. He was about 70 at the time, and for lack of anything better to do, he decided to watch us paint. Bringing along a lawn chair and a cooler of beer, he planted himself in the middle of the pool, and commenced to provide commentary on the quality of our work.

For whatever reason, we had given absolutely no thought to the toxicity of the paint fumes. Of course, in short order, the stale air laying in the pool stunk highly of paint. Still no clue. Dad's remarks began to be more and more whimsical and humorous. The four of us working started finding his remarks ever more hilarious. Soon, we ourselves were singing, laughing, and having a jolly good ol' time slapping that ol' paint on that ol' pool! Now and then we'd stop and have one of Dad's beers. This was turning into one big PARTY, and before long nobody could work without giggling. Dad was almost falling out of his chair guffawing at nearly everything.

Wait a minute, I finally thought, ...this isn't like Dad, even when he's drunk! And he'd probably only had a couple of the beers. The bulb finally lit! It was the paint! Chances are, if we stay here any longer, we'll probably go into the next stage of the effect of the fumes — most likely unconsciousness! We clambered out of the pool and helped Dad up the ladder. Within a few minutes all of the humor went out of the "party," and we all began to have headaches. While these diminished within an hour or so, we realized that we'd done a pretty stupid thing, and might well have been lucky to have lived through it!

We waited until the next day to resume painting, and this time set up a number of fans and blowers to carry the fumes away.

For his part, Dad only said "Well, I'm not going to watch you paint the pool the *next* time you do it!"

The Up-side Down Moray Eel

My friend Jack Porter and I were diving on the wreck of the *Benwood,* off Key Largo, in Florida. The wreck was sitting upright on the bottom, at a depth of about 50 feet. All of the ship's superstructure was gone, and numerous holes were rusted in her deck plating. Fish of all kinds abounded in and around the wreck, and Jack and I were after some pictures. I was particularly hoping for some shots of some juvenile spotted drum fish, which are spectacularly plumed. These were to be found mostly in the darker nooks and crannies in the wreckage, and I had asked Jack to be alert for them. He peered down into a hole about the size of a pumpkin in the deck, then excitedly waved me over to look.

"Great," I thought, "he's found a drum fish for me!"

I looked into the hole and saw nothing. I stuck my head completely into the hole, and still saw nothing in the dim light. I was now in the hole up to my shoulders, with my body vertical. I slowly started a 360 turn, hoping to spot what Jack had seen. At almost exactly half way through my rotation I found myself eyeball-to-eyeball, face-to-face, six inches away from a moray eel whose head was as big as mine!

Paralyzed for only an instant by the shock of it, I jerked my head from the hole. I turned immediately to Jack, who was giving me body language innocently spelling out "Well... did you see it?"

He has always claimed that the "it" he saw there *was* a drum fish. I've always claimed he set me up for that eel. He says he never saw the Moray. I say it was an ornery trick. Only he really knows, and he has never confessed!

You Can't Eat 'em If You Can't Catch 'em

In an earlier chapter, I described some divers who proved quite inept on their first dive for Florida lobsters. Here, I'll describe another guy who was the exact opposite.

I had scuba trained and certified my neighbor and good friend Dave Stemen. He and his wife sometimes accompanied us on Florida vacations that weren't primarily scuba trips. The first such trip after the completion of his course, we did plan to make a few dives, naturally. Dave was a lobster LOVER! Back home, in Ohio, one can buy lobster, or order it in a restaurant. Most are frozen, and not all that good, to me. Some are shipped in live, but even they aren't near as good as the lobster that goes in the pot within hours of when it is pulled from the ocean. But Dave was crazy about even the lobsters he ate in Ohio. I couldn't wait to have him sample freshly-caught lobster. I was sure that he'd be transported into gourmet's euphoria! Of course, he, too, was anxious to find out if there was as much difference as I claimed.

Over a promising lobster area I anchored the little 15-foot outboard boat we had trailered to Florida. This was to be only the second salt water dive Dave was to make, and he was still a bit apprehensive. Yet, his mouth was watering for those fresh lobsters, and he willed himself to prepare to dive. At the last minute, we realized we had brought along only one pair of gloves. "No problem," says I, "you wear one glove, Dave, and I'll wear the other. It only takes one hand to grab a lobster!" He took the right glove and I took the left, and down we went. I had the bag into which we'd place our catch. Lobsters were everywhere! Like a man gone mad, Dave started grabbing them, with both hands! In short order, he, alone, filled our bag. This guy had the knack! By the time we got back into the boat, however, he also had a bloody left hand. He couldn't have cared less about his wounds, and I've never seen a prouder look on a diver's face! That evening, he and his wife found out what lobster REALLY tastes like, and we ate ourselves silly. "You know," Dave said, "not only is this lobster ten times more tasty than what we get in Ohio, but I'll bet it tastes even better when you've caught the lobsters yourself!" I had to agree.

Memories, Memories

Through the years, I've tended to keep notes on many of the unusual things I've experienced in diving. Drawing on those notes for this book, I've probably written on less than 10% of them. Truly, it's been a blast! I feel sure that if hunting were my passion, instead, or any one of a number of other "hobbies," I'd likewise have collected many stories. But I remain convinced that it is the travel and the characters inherent to diving that make it so unique. If I had it all to do over, I'd travel exactly the same path. (Well... maybe I wouldn't paint the pool again, with no fans..)

13

Is Scuba for You?

While more than seven million folks have already been trained in scuba diving, that represents less than 3% of the country's population. Millions of other people are perfectly capable of enjoying the sport, and many have indicated, on public polls, that they would like to participate. Yet, many hesitate to take the first steps of learning more about it. Even those who *do* take the time to make contact with a dive shop or training organization, only a very small percentage proceed. This is borne out in our own experience, in that less than 10% of the callers who ask about scuba classes actually sign in.

The reasons for this reluctance are very clear. A number of wide-spread misconceptions exist about diving. Why? For one reason, it seems to be the tendency of many divers, even though they are true enthusiasts themselves, to exaggerate the negatives when relating to others the aspects of the sport. For instance, in our classes, we commonly have students comment to us, part way through their course, that they had understood from others that they would be required to do all sorts of impossible things. Some of their questions have included "When do we have to do the 200-yard underwater breath-holding swim?" This often makes us wonder how in the world they had the nerve to sign into a class if they thought all of these terrible exercises were actually going to be required. Then, too, many of the simple assignments that are the norm can easily sound impossible, when taken out of context. For example, a former student might tell a prospective novice that during the class it will be a requirement to place all of the diver's equipment on the bottom of the pool, dive to it, and don it under water. To the prospective student, this may well sound utterly impossible. He can't picture himself ever being able to do such a thing, so he decides not to sign in. In reality, this particular exercise takes place quite late in the course, after the students have practiced the skills leading up to it, and find themselves well able to handle it most proficiently and comfortably. And so forth.

Additionally, the media tends to over-dramatize potential diving hazards. Some TV shows and movies, for example, make it seem that the diver faces danger at every turn, and that diving is a constant effort to survive. Even documentary programs on diving often tend to inject an undue element of risk and danger. Life-threatening situations, it seems, are what "sells." Even scientific or research diving films often seem designed to cast the divers as "heroes" willing to risk their lives for science. Even when my own films, which contain no such content, are shown, I often hear gasps from the audience, as they misconstrue a perfectly safe scenario as being somehow "dangerous."

In general, 18 common misconceptions keep people from looking further into scuba diving. These are:

1. Diving is dangerous, and only for those who are willing to risk death or injury. It is a "dare-devil" sport.

2. Learning how to scuba dive, that is, taking a scuba course, is a difficult and/or harrowing undertaking.

3. One must be an athlete to scuba dive.

4. Scuba diving is prohibitively expensive.

5. There are too few places to scuba dive.

6. Scuba diving involves the ability to withstand great pressures on the body.

7. One must be an accomplished and talented swimmer to scuba dive.

8. Scuba diving is exhausting and physically demanding.

9. Scuba diving is mostly for young adults.

10. Scuba training is not readily available, and/or is too costly.

11. I have a medical condition that would likely preclude diving (and/or "my ears hurt when I go under water, so I could never scuba dive").

12. There are too many dangerous creatures under water, and I would be too nervous.

13. I can't hold my breath long enough to be a scuba diver.

14. I tried scuba diving one time (at a resort, or with a friend), and found it very intimidating.

15. It would be very cumbersome to have to wear all of that equipment.

16. My peers (family, friends, neighbors, etc.) would think me foolish to undertake such an activity.

17. I might like diving, but I would never want to go as DEEP as I hear you must go.

18. I've heard about the "bends," "rapture of the deep," "air embolism," and some of the other maladies and injuries that befall divers. I don't want that much risk in my life.

I will address these misconceptions, one by one, and provide objective answers to them. First understand, however, that I do not at all contend that diving is for everyone. At the end of this chapter, I will list those factors and characteristics that should preclude diving for certain individuals.

1. "Diving is dangerous — a dare-devil sport!"

Most any outdoor sport, and even active indoor ones, can sometimes result in injury. To rank scuba diving as hazardous as sky diving, motorcycle racing, and other high-risk activities is to not know the facts. To put it in perspective, having accompanied hundreds of divers on thousands of dives all around the world, I have never seen an instance of a serious diving injury. In fact, the worst problems I've seen are bumps and bruises from hauling equipment, and sunburn experienced by "snow birds" who want to take a quick tan back home.

As to "attacks" by sea life, people swimming on the surface of salt water are far more prone to such than are divers under water. Your odds of being bitten by a shark, in reality, are far, far less than of being struck by lightning in your own back yard.

Certainly, divers love to tell "war stories," and films often are based on the many "horrors" that abound in the underwater world. In real life, however, one would have to use a great deal of imagination to simulate a threatening situation. Indeed, one need not be "brave" to face the hazards of the dreaded "bends," embolisms, pressure "squeezes," etc. that are so often magnified out of proportion by the media. Modern training and equipment enable one to completely eliminate these hazards. Once a diver, of course, you are perfectly free to cast yourself in the image of a daring and fearless adventurer, if you wish. If you do so, take care not to ignore the fact that a lot of little old grandmothers do a lot of diving, too!

2. "Learning how to scuba dive sounds difficult!"

In the second paragraph of this chapter I related how divers far too often relate the difficulties of scuba training. Hey — they did it, and they are proud of it, and they are going to make it sound like a real accomplishment! Well, it is an accomplishment, but not nearly of the difficulty it is so often made out to be.

Many, many students, at the end of a scuba course, relate that the class itself was so enjoyable that they'd "like to do it again!" Modern scuba courses are conducted in a well-thought-out progressive manner. The early basic skills are easy to master. Each of these skills make the next one easy to learn. At no time is the student thrown into a water exercise with no preparation. And multi-media classroom teaching procedures make the learning process very effective, even for folks with a limited academic background.

We often hear folks say that they think they'd "kind of like to take a scuba course," but that they're "afraid I'd not do well and would be embarrassed." It should be kept in mind that most classes are made up of people from all walks of life, of many different ages, and of varying intellects. The vast majority are long "out of school," and it is common for them to wonder if they are still capable of study. The instructor will face a class that is comprised of many different backgrounds, and he or she will teach in a manner that will not "embarrass" anyone.

Sure, once you have completed a scuba course you can be proud of having accomplished something! It won't have been a "snap"! But it will not have been a course in rocket science, either, and you can be assured that it will have been designed for layman beginners. If someone tells you that, in their class, they had to do or learn this or that thing that sounds absolutely impossible to you, take this with several grains of salt. The average person can readily complete a scuba course, and enjoy the doing of it!

3. "One must be an athlete to scuba dive!"

In the early days of diving, when most instructors were influenced by the standards of military diving, only young, fit, adult males could pass the muster. And, only young, fit, adult males even tried, because diving was seen as a bold and gutsy thing to do. There wasn't that much leisurely exploring of warm-water coral reefs, back then. It was mostly a matter of cold-water shipwreck ravaging and/or other derring-do activities.

Today's modern equipment has removed much of the "hero" aspect of diving, and greatly enhanced its safety, as well as the diver's comfort. It is not so much a case of the physical standards having been "lowered," but more a matter of them having previously been unrealistic for civilian sport divers.

As far as physical strength or durability, the most demanding activity of diving is the loading and unloading of the equipment to and from the car and boat. And neat wheeled gadgets are available to make even that easier. Diving itself, conducted sensibly, requires a minimal amount of "strength." If you are going to enter competitive timed scuba events, naturally an athlete would have an edge. Or, if you intend to dive with "clunks" who you find yourself habitually towing back to the boat, you'd do well to have developed some adequate finning muscles. (Even then, modern flotation devices would make even "rescues" less strenuous.) But if you more wisely intend to limit your diving to the style that best suits your physical ability, you certainly don't need to be an Arnold Schwartzenegger! Again, remember those little old grandmothers diving. There's plenty of them. Even some great-grandmothers!

There is one caveat to this, however. Though one need not be a particularly strong swimmer to learn diving, it is important that one is capable of feeling comfortable in the water. The scuba course will start with some water activities designed to determine this. Those who may be capable in the water, but who have not been swimming much lately, will be given a opportunity to refresh their skills. These tests will not be on the competitive swimming level, by any means, and little heed will be given to the student's mastery of the various named swimming strokes. The instructor will, instead, be looking for an adequate "instinct" for the water. After all, if one has a natural fear of the water, and can't be comfortable there, scuba diving is not likely to be the sport of choice. On the other hand, it is natural enough to be "afraid" of the water if one has had a previous bad experience, or never been taught the most basic of water skills. Many people with these problems, but who still are intrigued by diving, have enrolled in swim classes first, and overcome these obstacles. My own wife, Alice, was one of these.

4. "Scuba diving is prohibitively expensive!"

"Prohibitively," of course, is relative. One person's fortune can be another person's pittance. To put this matter into some sort of perspective, here are the facts:

(A) A complete set of scuba equipment, including tank, wet suit, buoyancy control device, regulator, mask, fins, snorkel, and all the common accessories, will cost, brand-new, between about $1,200 to $2,000, depending on quality, brands, source, etc. Good used equipment can sometimes be purchased for half that amount. For warm water diving, where an exposure suit is not required, the cost will be substantially less.

(B) RENTING all of the above, per day of diving, will cost between $30 to $50.

(C) A complete scuba course, including the (local) field trip dives and most of the scuba equipment for use during the pool sessions, typically costs approximately $200 to $300, in most areas.

Of course, getting started, a new diver can own some of his or her own gear, and rent the balance.

Each dive one makes will have a cost of only about $2 to $3, for the re-filling of the scuba air cylinder.

Getting to the dive site, of course, and/or paying for the dive boat, will add more cost, but then one usually has to travel to engage in other sports as well.

Diving is not an especially cheap advocation. On the other hand, neither is golf, fishing, boating, or dozens of other things one can do. Once equipped, however, scuba diving is relatively inexpensive, in that the cost of the air fill is the only true expense, and that is minimal. And, many dive sites, unlike golf courses, have no usage fee.

As to the "investment" consideration, popular brands of diving equipment hold their value reasonably well (except for exposure suits, which depreciate in value much like used personal clothing.) Most equipment suppliers provide convenient payment plans for the purchase of equipment, to lessen the initial outlay. As to seeing the investment in terms of providing a "return" in recreation value, I have never heard a diver regret the cost of his or her outfitting. Never! Again, it may not be "cheap," but what a small price to pay for the capability of entering an entirely new world!

5. "There are too few places to scuba dive!"

Sure — there may well be no warm-water coral reefs in or near the state in which you live. Most of us have to travel quite a distance to get to them. But would you believe that guide books to dozens of local diving sites exist for nearly every state in the country? Even states that you might never guess would have dive-able waters. Of course, many of these sites are freshwater lakes, quarries, ponds, rivers, or springs, but many of them provide absolutely fascinating diving opportunities! Unless you are a diver, or know one, you likely would never know about such places. And, of course, anyone living within driving distance of any of the Great Lakes, and millions of us do, has abundant diving experiences at hand, in season.

Millions of others, of course, live within an easy drive from salt water diving. The coral reef diving off the southeastern coast of Florida, and in the Florida Keys, is virtually as good as any in most of the Caribbean, and all highways connect to there, as they do to the west coast and other areas of the east coast!

Some divers are perfectly content to dive only in fresh water, finding plenty to do and see. Fresh water diving, for most of us, is cold water diving. Modern exposure suits, "wet" or "dry," make this possible, and reasonably comfortable, almost year around. Others, usually termed "warm water divers," want only to dive in tropical salt water. They might not get to dive as often, but they seldom need to wear exposure suits when they do. Nothing at all wrong with that! A nice "balance" is enjoyed by others, who choose to dive both "locally," in fresh water, and to enjoy more "exotic" warm water diving on vacation trips. There is something to be said for this, because it keeps the diver active and practiced. Those "warm water divers" who only get to dive a week or two each year sometimes find it takes a few days each time to regain their comfort in the water.

"But where do you dive?!?!" is a question often asked of midwest divers. If you heard their answers, you might be astounded. Don't assume, as so many non-divers do, that one can only dive in the exotic places where Jacques Cousteau has opened up the frontier! The clearer the water, the more enjoyable the dive, for sure, but there are far more clear diveable fresh water sites available for diving than one might ever guess.

As to the accommodations for divers and diving that are now available around the world, one needs only pick up a current issue of any of the diving magazines. Literally thousands of diving facilities exist, and, believe me, they truly cater to the needs of the traveling diver! From scuba "camps" to the most luxurious resorts, from day-trip party boats to the fanciest of live-aboards, the diver can tailor a vacation to his or her tastes and budget.

And, speaking of vacations, what a wonderful enhancement scuba diving makes to a vacation trip to the tropics! I couldn't begin to count the number of folks that have remarked to me that they used to travel "just for sight-seeing," and then got into diving, and now so much more thoroughly enjoy their trips. What a difference a dive makes....

6. "One has to be able to withstand great pressures to scuba dive!"

Over and over, in movies and books, we're exposed to the concept that "great pressures" exist under water, and that the diver has to be trained to "stand" these pressures. I vividly remember a line from a popular film in which the diving hero says "I was so deep, and the crushing pressure so great, that I could hardly move my arms...." Poppycock! Not so! Hollywood garbage!

Yes, pressure increases as we penetrate deeper into the water. In fact, it increases about a half pound, per square inch, for every foot we descend. That doesn't sound like much, but it adds up! At a depth of only 33 feet, the surrounding pressure is already double what it is on the surface. At a depth of 100 feet, the surrounding pressure is nearly 60 pounds per square inch.

But does the diver actually FEEL this pressure? Absolutely not! All of the solid parts of our body, arms, legs, etc., sense absolutely no more pressure than at the surface. The increased pressure is transmitted perfectly evenly through our tissues, and no exterior pressure is felt whatsoever. Movement is in no way restricted. Breathing from scuba, the air in the lungs is automatically re-adjusted to the new surrounding pressure with each change of depth. No pressure is felt on the lungs. Pressure CAN be sometimes be felt on the ear drums, as one descends, but this sensation is easily eliminated each few feet as the diver "equalizes" the air in the inner ear with that in his lungs.

Being under pressure, and breathing air of elevated pressures does have certain effects on our body, and for these reasons we must limit our depth and time. This is easily done with the use of tables and/or modern instrumentation. Exceeding these limitations can have serious and damaging consequences, so the diver is trained how to stay well within these rules. At no time, however, does the diver experience the sensation of external pressures. Forget this!

7. "One must be an accomplished and talented swimmer to scuba dive!"

I addressed this under "Misconception #3." Comfortable and basically capable in the water, yes. An "accomplished swimmer," no. Don't sign up for a scuba course if the thought of being in the water panics you, for sure! But if you don't know an Australian Crawl from an English Backstroke, and can't do either one, don't worry about it!

If you sign up for a scuba course, and haven't been in the water for an extended period of time, you'll likely benefit from a bit of swimming before the course starts, if you have the opportunity. If this doesn't work out, simply tell your instructor that your skills may be a little rusty. He'll work with you.

8. "Scuba diving is exhausting and physically demanding!"

Yes it is, if it is your intent to enter competitive events or set new underwater distance swimming records!

If, instead, like most of us, your diving is recreational and explorative, you should never find a dive to be tiring, and should never return to your exit point huffing and puffing. If this happens, you've likely done something dumb. In reality, nearly every recreational diver I've known has described their diving to be exhilarating and invigorating.

Sure, you'll likely sleep well after a day of diving, and you will probably have a big appetite for dinner. Diving is an outdoor sport, it's active, and it's not for serious couch potatoes. But it is not "HARD," and, conducted properly, it should never tax your stamina. On the other hand, if I had two diving buddies to choose from, and one was fit and one wasn't, I'd likely choose the fit one. (All other things being equal..) Why? Hey..., I just might choose this dive to do something really stupid, and need my buddy's help. That I am saying that diving is not strenuous is not an argument for letting yourself get out of shape! People who keep themselves in good shape live longer, are generally more fun to be around, and, *as a side benefit*, they tend to use less air.

Remember, too, that diver with one leg and one arm that I described in an earlier chapter? Lot's of handicapped folks enjoy diving, and dive safely. I've known some that had to be brought to the water's edge in a wheelchair! Certain medical problems can preclude diving, of course, but diving is definitely not only for fine physical specimens.

9. "Scuba diving is for young adults!"

Tell that to my 66 year old favorite diving buddy, or that great-grandma I often dive with! Or that 77 year old dude in Hawaii who dives almost every day!

Anyone of any age can dive, except those who are too immature to know and follow the safety rules, or are too old to remember them, or who are too infirm to render assistance to a needful diving buddy or who have contra-indicative medical problems.

10. "Scuba training is not readily available and/or is too costly!"

I spoke about the cost of diver training earlier. In my personal opinion, it is a good value. And, you can be sure that no one is getting rich from providing diving classes, as doing so is far too competitive, in most regions. From experience with hundreds of students, I feel sure that most don't regret the expenditure, if they can scrape the money together. With some courses, a payment plan is available, to make the payments a little more painless. And, sometimes an offer is available to provide two or more people enrolling at the same time with a discount on the course.

As to the availability of scuba training, this is becoming less and less of a problem as the sport grows. In an average city, one can choose from several facilities offering the training. Since a pool is required, the training might involve a bit of a drive (but probably not the 120 mile round trip that mine did!). If no classes are being advertised to the public in your area, call the YMCA, YWCA, or any dive store. Classes are usually held year around. Most are conducted in the evenings, to fit with most people's work schedule, but courses are sometimes also held on weekend days. If you are in a hurry for your training, or want to be in a smaller class of only folks you know, inquire about private classes, which are available through most dive stores, and can usually be custom-fit to your own schedule.

11. "I have a medical condition that likely would preclude diving!"

Certain medical conditions DO preclude diving. Heart problems, lung disease, and serious epilepsy are some of these.

Serious claustrophobia is also contra-indicative to diving, though many folks count themselves claustrophobic who really aren't. If you've ever had a mask on under water, without a feeling of panic, you're likely OK for diving, in this respect....

The conviction that "my ears hurt so bad when I dive under water that I could never be a scuba diver" is a factor that needlessly keeps a lot of people out of the sport. MY ears would hurt, too, when I descend into the water, if I don't handle the problem. Almost everyone's ears do, and it's really painful for some. But, it is entirely normal, and to be expected, and early in a diving class everyone is taught the simple techniques of totally eliminating this nuisance! It's easy! It's not a matter of simply learning how to stand the pain, but the learning of how to avoid it altogether. DON'T LET THIS KEEP YOU FROM LOOKING INTO DIVING! It's normal! You're not unique in this!

Others say "I'd like to learn scuba diving, but I had a ruptured ear drum when I was a child, so that means I can't dive." Almost certainly untrue. Many folks get ruptured ear drums sometime in their life, sometimes several times. In the vast majority of cases, these ruptures heal completely. (If you really think yours didn't, duck your head under water, hold your nose, and force air up through your eustachian tube into your inner ear. If bubbles come out of your ear, you have an un-healed hole in the drum. If not, it's healed, and would be no problem in scuba diving!)

Nearly all scuba courses will require that you first, or very early in the class, go to a physician for a complete medical exam, at your own expense, providing the instructor with a form signed by the physician attesting to his opinion that your medical condition and history is compatible with scuba diving. If, at this time, you have concerns about past injuries to your ears, or any other considerations that may not be immediately obvious to the doctor, you should make it a point to discuss these with him.

12. "There are too many dangerous creatures under water and I'm sure that I would be too nervous about them!"

My youngest daughter, an excellent swimmer and an ardent outdoors person, will not scuba dive because she "just couldn't be comfortable with all those live fish swimming near her..." In fact, she doesn't like to swim, or even wade, in waters known to contain "live things." No problem. I've never coaxed her to dive or ridiculed her for this phobia. Perhaps there are others who feel this way.

But as far as a fear of being attacked, wounded, or *eaten* by something in the water, I hope that my earlier remarks may have helped to assuage this apprehension. Certainly, there have been people who have been hurt by such as sharks and barracudas. In essentially every case, however, the circumstances were far outside the norm. In some cases, divers were spearing fish and made the mistake of "arguing" with a predator fish that wanted the game. Or, a swimmer on the surface was mistaken for an injured form of edible sea creature. Or, in sheer stupidity, the diver was threatening or antagonizing the shark or barracuda.

The worst fish attack I've ever suffered was in a freshwater quarry. I was with a student, and stopped him in an area where several yellow perch could be observed. We weren't wearing wet suits. I showed the student how the perch would actually come up and nibble peeling sunburned skin from my forearm. Fascinating! Then, suddenly, one of them got under my life vest and grabbed my nipple. Hard! He obviously intended to take it home with him! I shooed him away, but he actually drew blood! Now, this doesn't make a very impressive war story. Non-divers, when learning that one is a diver, usually ask these three questions: "How deep have you been?," "Have you ever had the bends?," and "Have you ever been attacked by a shark?" To the latter, I seldom admit that the closest I ever came to a shark attack was a nipple nip by a one pound perch!

Again, injuries and attacks from underwater inhabitants are so rare as to not be a consideration of diving. Sure, barracuda have the unsettling habit of staring balefully at one, while sometimes "gnashing" their teeth and hanging motionless in the water, usually behind you. You might swear, sometimes, that you have heard one growl. (They don't.) And, large barracuda, indeed, have a mouthful of sharp teeth that would put a doberman to shame. But they simply don't attack divers. We used to say "don't wear anything yum-yum yellow in salt water, or anything that shines like a fishing lure, or a 'cuda might make a strike at it!" It doesn't happen. On the other hand, it wouldn't be wise to kick the neighbor's friendly dobe, would it?

In reality, it is a real treat to observe these creatures in their own environment. Don't turn away from diving because of a fear of them. Get down there and SEE 'em! You'll be glad you did, and if you get eaten, we'll refund the price of this book!

13. "I can't hold my breath long enough to be a scuba diver!"

In scuba diving, no breath-holding is involved. During a scuba course, the student will first learn the basics of skin diving (snorkeling), which does include some exercises in breath-holding, and, once the training with the scuba equipment starts, some of these training procedures will involve a certain amount of breath-holding. After that, the diver may well experience an entire life time of diving with no breath-holding involved. Most scuba divers, however, often enjoy snorkeling, in between scuba dives, especially in reasonably shallow areas of interest. Snorkeling, of course, can be conducted entirely on the surface, breathing through the snorkel, though the diver will likely find occasion to submerge for brief periods to observe something more closely under water. (He is then "skin diving.") Obviously, the better one's breath-hold ability, the more time can be spent under water in this activity.

So, while scuba diving doesn't require breath-holding, it is to the diver's (and student's), advantage to improve their breath-holding time. A vast improvement in this capability can be expected simply with the practice involved in a scuba or skin diving course. Furthermore, the instructor will demonstrate certain practices which will greatly aid in this. And, if the individual further practices this a few times between each class session, even further improvement can be made. Typically, a person who, with the first effort, might only be able to stop breathing for, say, 15 seconds, can quite readily improve this to a full minute, or even more!

In any case, an otherwise proficient scuba student with a breath-hold capability of at least 20-30 seconds should readily be able to complete any underwater assignments during a typical scuba course.

14. "I tried scuba diving one time and found it very intimidating!"

You and hundreds of other people! The majority of which were put into a diving situation with far too little preparation. Far too often, enthusiastic new divers are eager to show their friends how easy and enjoyable their new sport is, and willingly let someone untrained "try out" their gear. Without adequate indoctrination, this almost always results in a negative experience, and a real turn-off for someone who might otherwise have truly enjoyed diving. Don't do this — ever! Aside from the probability of disappointing someone, it can also be quite dangerous!

Some folks also participate in an "Introduction to Scuba" dive while a guest at a hotel or resort in a tropical area. "Sounds like a fun thing to do, and the man said it would be completely safe!" Perhaps, indeed, the experiment is conducted in a reasonably safe manner. (Some aren't!) Still, the person experiencing this with no skill preparation leading up to it is likely to feel a great deal of apprehension, and perhaps even encounter a distressing situation. Some people have been "turned on" to diving in this manner, but many more have come away from it convinced that diving was not for them. A novice simply cannot be adequately prepared for the underwater world with a simple top-side briefing. Modern courses are designed to create more and more confidence with each step of training. For most people, it does take a little time to feel at ease under water.

If you'd like to "sample" scuba before signing up for a class, many dive stores now make a "discover scuba" experience available, at little cost, in a swimming pool, under carefully controlled conditions. You won't see a coral reef, or any tropical fish, but the experience is much more likely to be a pleasant one!

Truly, it's not "hard" to learn how to breathe from a regulator. Anyone could master that in only a few minutes. Even a small child. It's the learning how to handle all of the equipment, enter and exit the water safely, manage one's motions, and how to avoid situations that could lead to emergencies that requires the time involved in a full class.

15. "It would be very cumbersome to have to wear all of that gear!"

To see a fully equipped scuba diver out of the water, one would naturally think this is a very "encumbered" sport. Indeed, an equipped diver out of the water is usually about as graceful as a cat wearing galoshes. Once under water, however, all of this heavy-looking gear becomes virtually weightless. Each piece of the equipment has a purpose, and its purpose is to aid the diver, and make him more comfortable. It is not a "drag" or an impediment. Someone might say, "Gee, it would be easy enough to fall into the water with all of that gear on, but how would one ever get back out?" In reality, the scuba student is taught many techniques that make it delightfully easy to re-board a boat. It need not be a "muscle affair" by any means.

True enough, it does take a certain amount of time to don all of the equipment, particularly if an exposure suit is required. This is why a lot of divers especially enjoy warm water diving, when the exposure suit is not required.

16. "My peers would think me foolish to undertake such an activity!"

Maybe you have the wrong peers! Several places in this book I've referred to a common denominator among most divers as being "individualistic." Are you? Chances are, once you're diving, and your friends, peers, and family see the new sparkle in your life, they'll be glad you got involved!

As to how "foolish" diving might be, consider this: would your peers consider you wiser if you instead took up motorcycling? Sky diving? Snow boarding? Bull riding? Water ski jumping? Drugs? And, of course, there is always the chance that you might surface, from any one dive, with that golden chalice! Few other sports hold out that possibility!

17. "I might like diving, but I don't want to go DEEP!"

So don't go deep! There is very little reason to do so! By far, the very best diving, scenery, and color is to be found at depths of 10 to 60 feet. Much beyond that, colors fade, fewer forms of sea life are seen, and the diver's air supply lasts for a far shorter period of time. We've taken dozens of diving trips to destinations all around the world, and most often no one has reason to go any deeper than that.

It is a fallacy that "deeper is better." Once you're diving, friends will ask you "How deep have you been?" They assume that it is a measure of your skill and ability to be able to go deeper and deeper. If you answer "I've been to 60 feet!," and they know of another diver who answered "100 feet!" they'll just naturally think that other diver is "better." Let them! Or lie to them (your C-card gives you a license to lie about depth). Say "Hey —

I've been to 300 feet!" You don't have to go there to say it and, what the hey, if it impresses them, so be it!

The fact is that there is usually less to see at 100 feet than at 40 feet, and less light and less color, and far less time to stay there. And, it doesn't take "skill" or strength to go deep. It's easy. But it's dumb. Tell *me* you've been to 300 feet, and I'll count you *dumber* than the guy who's never been beyond 60 feet! And, because of nitrogen narcosis, the deeper you go, the dumber you get. And deep isn't the right place to be dumb!

Sometimes, when we're anchored on a reef that drops from shallow to deep, a diver will request permission to go to a depth of 100 feet, "Just to say I've been there!" Perhaps, under the right conditions, I'll accompany him there once, briefly, just to prove to him that it isn't worth going back. More often, however, I'll tell him to "*go* ahead and SAY it! You don't have to go there to say it. And, as a matter of fact, I'll give you a signed note testifying that you were there, without even needing to actually do it!" Is that fair?

Haven't *I* ever been deeper than 60 feet? Sure, a few times. But only if there was a shipwreck or something else there worth seeing or getting that couldn't be seen or had at a shallower depth. For one thing, I enjoy my dives lasting as long as possible. If what I am looking at or photographing is less than 33 feet deep, great! At that depth, I don't even need to think about time limits or avoiding the bends. And, my air lasts much longer! An example: if a tank of air lasted a diver 60 minutes near the surface, it would last only 30 minutes at a depth of 33 feet, 15 minutes at a depth of 99 feet, and only about 7 minutes at a depth of 200 feet. It would take nearly 7 minutes to get to a depth of 200 feet, and back! Why bother?

Don't worry about DEEP! If something would prevent me from ever going deeper than 30 feet the rest of my life, it wouldn't slow down my diving a bit!

18. "I've heard about the 'bends,' 'rapture of the deep,' 'air embolism,' etc., and don't want to expose myself to those risks!"

There are certain risks in any undertaking. Even driving an automobile is hazardous, or fatal, if one disobeys certain rules of logic and safety. Gawk at something beside the road while you're driving, allow your car to drift a couple of feet left of the center line, and in an instant you can be dead or horribly crippled for the rest of your life!

Frankly, these "exotic" hazards of diving have been vastly over-played in the dramatizing of such as movies and TV shows. When one makes a diving film, it is almost certainly going to be an "adventure" film. Adventure films have villains and dangers, and the hero is exposed to all types of life-threatening situations. Real diving isn't like that, except perhaps for idiots. I've known hundreds of divers who have logged many thousands of underwater hours without getting "bent," "embolized," etc. It is simply a matter of learning how to keep yourself from violating the basic safety rules of diving. These hazards are not mysterious or random, and are easily avoided.

These dangers are, however, hard and fast factors of physics and physiology, and no diver becomes immune to them. Exceed the limitations of time and depth, or the appropriate rate of ascent, and you become a candidate for a very real problem. Part of a scuba training course is the learning of the safety parameters of diving, just as an automobile driving course involves learning how to avoid highway accidents. Be a dare-devil driver, and you'll get hurt. Likewise with scuba. At least with scuba, no other driver will ever come across the center line and nail YOU!

Then, what would, or should, keep me from diving?

(A) Diving is not for everyone. Those who have a real fear of the water won't enjoy it. After all, diving is water — all the way!

(B) And, certain medical conditions, such as mentioned earlier, will preclude diving.

(C) Diving is clearly an active OUTDOOR sport. Folks who don't like or appreciate the outdoors probably won't find much pleasure in diving.

(D) Divers sometimes get COLD. Getting cold, in itself, isn't much pleasure for anyone, but enthusiastic divers learn to tolerate it. Even long dives in tropical water, especially on cool cloudy days topside, can eventually chill a diver. Anyone with a drastic aversion to chilling might not find an adequate balance with the pleasures of diving.

(E) Diving is a buddy sport. We don't dive alone. Buddies count on each other. "Individualistic" we may be, but we still dive in twos. Anyone determined to do everything ALONE shouldn't get into diving. (Or tennis, for that matter...)

(F) People who are careless about the environment shouldn't get into diving. Maybe they'd enjoy it, but we don't need folks tearing up the reefs, polluting the water, and generally trashing up the planet! Gather up these types and let them colonize Mars, and ruin that!

(G) "Show offs" shouldn't get into diving. They'll ruin our safety record, and our image. Let them go into the woods and color each other with paint balls.

(H) Youngsters under the age of 12 shouldn't scuba dive, nor anyone of any age who is not mature enough to perform as a competent dive buddy. We are sometimes asked to train very young people with the disclaimer that "a parent will always be diving with them, and looking out for them." Fine. But who does this then leave to "look out" for the parent?

So... all of the above negatives might eliminate some 30% of the U. S. population. Three percent of the population is already diving. That means that the remaining 167,500,000 other citizens are perfectly suited to participate in and thoroughly enjoy diving! Go for it! Whoa..., wait a minute. There wouldn't be enough water to go around! O.K., then, *half* of you stay home and watch television instead. Which half will YOU be in?